THE EMERGENCY–DISASTER SURVIVAL GUIDEBOOK

by

Doug King

WARNING

ISBN: 1-883736-10-2

For distributor information write to:
ABC Preparedness Co.
PO Box 795
Sandy UT 84091

Excellent Book

Table of Contents

SCARS Coupon
20% discount for worlds best fighting system,
(very simple, anyone can learn) See page 94–95

See the bottom of page 100 for companies that sell disaster supplies

1

Personal Safety in Cities and Automobiles

Read through this chapter for important safety tips and information. We also highly recommend that you purchase the book *The Truth About Self-Protection*, Bantam Books®.

Safety in the City

- Some states will allow you to get a driver's license using a post office box for an address. This way, people will not know your home address.

- Get an unlisted phone number. Otherwise, criminals can use your phone number to find your address in a special directory.

- Never open the door to a stranger.

- Get a dog. A large, barking dog is an excellent deterrent to discourage thieves, muggers, or burglars.

- Avoid expensive looking jewelry and cars — they attract thieves.

- Keep most of your money in a money belt when walking around the city.

- Never walk by yourself — especially at night. If you have no choice but to walk alone, bring a big dog along.

- If you are caught in a bad area — do not act scared. Criminals can sense fear and vulnerability. If you are in serious trouble, run while yelling, "Help, Fire!"

- When you are caught in a dangerous area, go to the nearest police station, fire station, hospital, large hotel lobby, restaurant, or other safe area with lots of people. You can then safely call for help or for someone to pick you up.

- Have a CB radio installed in your vehicle or carry a portable CB radio. You can use Channel 9 to call for help. The police regularly monitor transmissions on Channel 9.

- If you are able to get to a pay telephone and don't have any coins, you can usually still dial a friend or relative for help by using the following toll-free numbers:

  1-800-OPERATOR (1-800-673-7286) or

  1-800-COLLECT (1-800-265-5328)

- To have money wired, call 1-800-CALL-CASH (1-800-225-5227)

- If you are being chased, yell "Fire!" Don't yell "Help!" And do not stop until you are in a safe area.

- Avoid bad areas completely! The local police can tell you how dangerous a particular area is.

√ Never stop and talk to people on the street. Criminals often ask you for change or directions, then pull you out of sight where they rob or assault you.

√ Always carry travelers checks ($50 or more, American Express® is preferred) for emergencies.

• See the chapter on *PREPARATION TASKS and SKILLS* (page 41) for information about joining an auto club such as Automobile Association of America (AAA®) and/or adding towing coverage to your car insurance.

• You can reduce crime in your area by up to 50% by starting or joining one of these national groups. Call your local police or write to:

NEIGHBORHOOD WATCH
LA Police Department, Devonshire Division
10250 Etiwanda Avenue, Northridge, CA 91325

NEIGHBORHOOD ANTI-CRIME CENTER
Citizen's Committee for NYC
305 Seventh Avenue, 15th Floor, New York City, NY 10001

Safety in your Automobile

• Riding in a car with a proven safety record could save your life in an accident. The Volvo 240 is considered the safest car. Plymouth Voyager, Dodge Caravan, Ford Aerostar, Buick Riviera, Jaguar XJ6 and the Mercedes 190D are also excellent.

√ Do not drink and drive! Half of all auto accidents are alcohol related.

• Speeding contributes to many accidents. Don't go more than 5 miles per hour over the speed limit. And always remember to wear a seatbelt.

√ Stay in the open areas while driving. Don't drive with the packs of cars. Keep one car space of distance between you and the car in front of you for every 10 miles per hour of speed you are traveling.

√ Keep a half tank of gas or more in your car at all times. Full tank is best.

√ Keep emergency supplies in your car: water, food, flares, light sticks, tools, first aid kit, winter clothes and boots, blankets, etc. Also see *SUPPLIES - ADVANCE PREPARATION* chapter (page 68) for more details.

√ A cellular phone can be carried in a car for emergencies. You can also purchase a portable CB radio that can be used almost anywhere (the signal may not reach as far a distance as a permanently installed CB). The police listen to Channel 9. You can radio for a tow truck or other help.

√ If you are able to get to a pay telephone and don't have any coins, you can usually still dial a friend or relative for help by using the following toll free numbers:

1-800-OPERATOR (1-800-673-7286)

1-800-COLLECT (1-800-265-5328)

√ Only park in well-lit areas. ~~If possible, park directly underneath a street light.~~

√ Have your keys in your hand when you walk to your car.

√ Always look in the back seat, underneath and around your car before getting inside it.

√ Avoid very expensive cars (they attract criminals).

√ Stay out of bad areas — especially at night. The local police can tell you how dangerous a particular area is.

√ Keep up the maintenance on your car (especially the tires, hoses and belts).

√ Before leaving on a trip, have a safety check done on your car by a certified mechanic.

√ Keep all valuables including purses and briefcases out of sight at all times both while driving as well as when parked.

√ Lock the windows and the doors while driving as well as when the car is parked.

√ If your car is in a minor accident in an unsafe, deserted, or otherwise dangerous area — don't get out of your car or roll the window down. *CAUTION: Only drive the car after an accident if it is safe to do so. Motion to the other driver to follow you to a safe, well-lit place with plenty of people. WARNING: Car thieves and other criminals may create accidents on purpose so that they can stop you and steal your car with the keys still in it. Many criminals also cause minor fender-bender type accidents to set up an unwitting victim for robbery.*

√ Do not stop and ask directions in an area you think is unsafe. Wait until you are in a better area. Good places to stop to ask for directions or help include a police station, fire station, hospital, large hotel, or restaurant.

• See the chapter on *GENERAL PREPARATION TASKS and SKILLS* (page 41) for information about joining an auto club such as AAA® and/or add towing coverage to your car insurance.

Protecting Children from Abductors/Molesters

√ Teach young children never to open the house door for, or talk with, strangers, never go anywhere with strangers and never to help or get into unknown cars with strangers. If a stranger asks the child to get into a car or to go anywhere with the stranger, the child should run away while screaming and report the incident immediately. Teach your children that no person is allowed to touch their private parts. Teach children to run away while screaming and immediately find and tell a parent, trusted adult, school teachers, etc., if anyone tries to molest or abduct them. After instructing them, test their knowledge and response by playing the role of the "stranger" without actually touching them. Repeat training until they have the correct response. Test and train children regularly.

Note: Contact the local police department for information, groups and suggestions on protecting children.

2

Clothing for Cold and Hot Weather

Clothing must be kept dry to keep you warm. Gloves, scarves and hats are very important because more than 50% of your body heat is lost through your head, neck and hands. If you have the correct clothing, you can survive — at least temporarily — in very cold temperatures without a shelter.

√ Wool is the best type of non-specialized clothing for cold weather. Wool stays warm even when wet.

√ Wear several layers of dry clothing to trap the warm air that your body generates.

√ A winter jacket or coat that sheds rain is a very important item to have. The type of materials or fabrics that "breathe" (lets the moisture that the body produces escape) while still keeping you warm is ideal.

Emergency (Improvised) Cold Weather Bag

√ Find a large piece of plastic or cloth.

√ Fold it and tie it up so it is the shape of a sleeping bag.

√ Fill it with dry insulating material such as dry grass, shredded paper, foam from the inside of your car seats (which works very well), or cattail fluff.

√ Then use this "bag" as a blanket or a sleeping bag.

Emergency Cold Weather Clothing

Keep towels on hand. Towels make excellent clothes. Tie them around your arms, legs, head, etc. Newspaper can also be used in the same manner.

For more information on what to do in cold/stormy weather, read the following chapters: *HEAT and TEMPERATURE: How to GET or STAY WARM* (page 20) (which includes sections on Frostbite and Hypothermia), *SHELTER* (page 63), *SHOES* (page 66), and *WINTER STORMS* (page 83).

Clothing for Hot and Sunny Desert Areas

A hat and sunblock are extremely important to wear in hot, sunny areas. It is usually best to wear clothing made from lightweight cotton-type materials that are white (or a very light color): dark colors absorb heat while light colors reflect heat. Also, cover your head with a white cloth if you don't have a hat. This works better than wearing nothing on your head. See the chapter on SHOES for more information for emergency sandals and the illustration of a Desert Shelter on page 50.

3

Direction Finding

When people are lost, they tend to travel in circles because it is easy to lose one's bearings without permanent landmarks to follow. To avoid this sometimes costly mistake:

√ Decide which direction to travel.

- Use one of the methods shown in the illustrations on pages 52 and 53 to find North, South, East and West.

√ Pick out the most distant object or landmark in that direction.

√ Travel straight toward that chosen landmark.

4

Earthquake

Earthquakes are sudden and can be very powerful. It is very important that you are fully prepared before an earthquake strikes.

Advance Preparation

Buy the Standard Supplies that are vital for any disaster. These are listed in the *SUPPLIES - ADVANCE PREPARATION* (page 68) chapter under the heading Standard Disaster Supplies. It is very important to read the entire chapter. Most of these items can be purchased from the company listed at the bottom of page 100.

In addition to the Standard Disaster Supplies, you should stock the following Special Supplies in case of an earthquake:

☑ Store a sturdy pair of shoes and leather work gloves under each bed.

☑ Store a propane or wood burning stove and fuel for heating.

☑ Store a portable grill (or hibachi) and charcoal to use for cooking outdoors.

☐ Store regular car antifreeze (or the nontoxic kind used for RVs if available) for winterizing toilets, plumbing, etc. See Winterize and Go in the POWER FAILURE - BLACKOUT chapter (page 39).

☑ Store a roll of plastic (12 feet by 100 feet long.)

Preparation Tasks and Skills

☐ Learn in advance where your community evacuation routes are.

☐ Learn in advance the location of your local disaster shelters.

☐ Learn what the bomb/disaster warning sirens sound like.

☐ Read through this entire book so that you are more familiar with various types of disasters and the advance preparation needed.

☐ Read the chapter GENERAL PREPARATION TASKS and DISASTER DRILLS (page 41) for additional information on being more prepared for emergencies and Disaster Drill (page 42) to have with your family.

☐ Purchase earthquake insurance. Call your insurance agent for details.

☐ Anchor heavy bookshelves (or other items that may fall over) to the wall. Also, anchor valuable items (computers, etc.) to tables with velcro®, straps, or similar materials to prevent damage. Install childproof latches on all cabinet doors to keep them from opening during an earthquake.

☐ TIE DOWN THE WATER HEATER: Attach it to the wall with two or three strips of metal plumber's tape or secure it with a commercial fastening kit. During an earthquake, water heaters fall over which ruptures gas lines and causes fires.

Have
☑ ~~ATTACH~~ A PERMANENT SHUT-OFF WRENCH TO THE OUTSIDE GAS METER: This way the gas can be shut off from the exterior of the house. ~~These wrenches are available from the suppliers listed at the bottom of page 100.~~

During an Earthquake

✓ Remember that broken glass, falling objects and fires (from damaged gas lines, etc.) are what hurt the most people in an earthquake.

✓ If you are INDOORS — stay indoors! Get under a sturdy table, desk, or an interior doorway. Stay away from fireplaces, windows, tall bookcases and anything that could fall on you. *NOTE: Do not get under a section of floor that has a refrigerator or other heavy object above. These could fall through the floor and land on top of you.*

✓ If you are OUTDOORS — get out in an open area away from buildings, power lines, trees and anything that might fall on you.

After the Earthquake

WARNING: If you live in a coastal area, beware of severe tidal waves that can follow an earthquake. Stay tuned to the radio or TV for warnings, instructions, and information.

☑ Turn on the radio and TV for official instructions and information.

☑ Check for injured or trapped persons and render first aid or help as needed.

☑ Get ready for after shocks since they often take place following an earthquake.

☑ Do not enter any damaged buildings or houses. Inspect your home for damage. If your dwelling place is unsafe, go to the nearest shelter or to the home of a friend or relative in a safer location. Take a 72-Hour Emergency Kit with you if it is safe to get it from its storage place (see page 68).

☑ Don't use matches, lighters, candles, or have any open flames indoors until you are positive that there are no gas leaks or flammable chemical spills in the area. If you use a flashlight, turn it on outside the house before bringing it inside the house. The safest light source to use is a chemical light stick that you bend to activate.

☑ Turn off the electricity, water and gas (from the outside of the house) if they are damaged or leaking. Read the chapter TURNING OFF the GAS, ELECTRICITY, and WATER for details (page 78).

☑ Clean up spilled chemicals or medicines if it is safe to do so.

☑ Don't drink the faucet water until you read page 79.

☑ If you must cook, do it outdoors on a grill or in a pit in the ground.

☐ Don't use the telephone unless it is an emergency.

☑ Contact your insurance agent if there are injuries or property damage.

5

Evacuation

During an emergency, listen to a portable radio for instructions on where to go and when to evacuate. When you evacuate, take the exact route that the authorities tell you to take. Don't take any shortcuts. Your life may depend on how well and how quickly you follow the instructions and directions of authorities.

Prepare in advance to take with you the supplies listed below if there is time. Read the SUPPLIES - ADVANCE PREPARATION chapter (page 68) for the Regular and complete List of Disaster Supplies.

- ☑ 72-Hour Emergency Kit
- ☑ Food
- ☑ Water
- ☑ Car gas tank filled with gas.
- ☑ First aid kit and first aid instructions.
- ☑ Blankets and winter clothes.
- ☑ Money in one dollar bills.
- ☑ Portable heater and fuel.
- ☑ Shelter (tent, tarp, plastic)

If the authorities on the TV or radio advise you to (and you have time), you can prepare your house for the disaster by doing some or all of the following tasks.

NOTE: *Only do the tasks that the authorities tell you to do or important tasks that there is sufficient time to safely complete.*

- ☑ Turn off the main water shut off valve (see page 56).
- ☑ Turn off the master electricity circuit breaker or fuse box (see page 57).
- ☑ Turn off the gas shut off valve where it enters the house (see page 58).
- ☐ Board up windows and doors in the case of hurricanes and tornadoes.

Read the chapter on TURNING OFF the GAS, ELECTRICITY and WATER for more complete details (page 78). Also, read other chapters for specific types of emergencies and disasters you may encounter.

6

How to Build a Fire

Knowing how to build a fire can be very important — especially in cold weather. Read through this entire chapter before you build a fire. REMEMBER: It takes patience to get a fire going — so don't give up!

Keep your fire small so you can get close for warmth. This is the way the Indians did it. When you build a large fire, your back gets cold while the front of you bakes. When you make a fire, stack a pile of rocks or wood on the opposite side of the fire so that the heat is reflected back towards you. *Note: After your fire dies down save the coals by banking them into a pile and covering with an inch of dirt. Later on you can uncover these coals and use them to start another fire.*

Automobile Tire Fire

- A spare tire from the trunk can be lit on fire with a flame (match, lighter, etc.). This can be used to keep you warm and also be seen as a signal for many miles. You can also take a tire off the car and burn it if necessary. Just remember, if you remove a tire from the car and burn it after you've burned your spare tire you will no longer be able to drive the car. **CAUTION:** Tire smoke is harmful, avoid breathing it. Also, once the tire is burning, it is almost impossible to put it out.

First Stage: Tinder and Starter Wood

This is the first and most important part of the fire. For tinder you can use dry shredded bark, fine dry grass, paper, or any fine dry material that will easily catch on fire.

- ✔ Tinder material should be finely shredded. To do this, you can rub it between your hands or pull it apart. The tinder should be the thickness of a sewing needle (or less if possible) when you are finished.
- ✔ After the tinder is shredded, place the tinder in a bundle about three inches in diameter and six inches long. Tie it at both ends. You can tie it with strips of green bark or green grass. This is called a tinder bundle. See the illustration on page 43.
- ✔ Next, gather up dry twigs and small sticks. Start with the thickness of a match stick and graduate up to about 1 inch thick.
- ✔ For your starter wood you can find dry, dead branches on the lower sections of trees. These are usually easy to find and are dry even when everything else is wet.

Starting and Maintaining Fires

Listed below are several methods of starting fires:

- ✔ Gas from the car gas tank can be poured on a rag, scrap of clothing, wood, etc., and used to start a fire. **WARNING: GAS IS EXTREMELY FLAMMABLE, only use a small amount (about a quarter cup.)**

✓ The easiest way to start a fire is to use matches or a pocket cigarette lighter. A car cigarette lighter also works well.

• Another method is to focus or concentrate the rays of the sun. Use a magnifying glass or a flashlight reflector that has a cigarette or some dry tinder placed in it — these should quickly catch on fire. See the illustration on page 44.

✓ If you happen to have steel wool and a battery (such as from a flashlight or camera), a fire can be easily started. Touch one end of the steel wool to the positive pole of the battery and the other end of the steel wool to the negative side of the battery. The steel wool will immediately catch on fire and you can then easily start your tinder bundle on fire.

• Another method to start a fire is to disconnect the positive battery cable on your car and touch it to the metal body of the car (as far away from the battery as possible). It will work even better if you attach a piece of thin wire (nonessential) to the battery cable first. This will produce sparks. Place the tinder on the spark until it catches on fire. *WARNING: Use extreme caution with this method — the gas fumes from the battery can catch on fire and cause the battery to explode. If you use this method, first place a cover over the battery. Remember to touch the battery cable to the metal body as far away from the battery as possible.*

• An even better way is to use a piece of hard steel and a rock that will produce sparks when they are struck together. Flint rock works very well. Before you try to produce a spark, make a tinder bundle as previously explained. Then strike the rocks (or steel and rocks) together. Make sure the sparks go down into the middle of your tinder bundle. You can purchase a flint and steel set from most camping or emergency supply stores. *NOTE: This method can take a long time — so be patient.*

✓ When a spark catches on the tinder, gently blow on the bundle until it catches on fire. Then place the lit tinder bundle under the pile of sticks as explained below.

✓ The best way to start a fire is with a tinder bundle as previously explained. Once the bundle catches on fire, place it underneath a pile of sticks that have been set up like a tepee. Gently blow on the bundle so that the wood sticks catch on fire. Make sure that the smaller sticks are near the bottom by the tinder bundle. See the illustration on page 43.

✓ The other method is to place the tinder on the ground and then light it. After the tinder bundle lights on fire, place tiny sticks (a few at a time) on the top of the lit tender bundle. As the sticks catch on fire, slowly add larger and larger sticks. Then finally add logs if they are available.

✓ Split logs burn better and put out a lot less smoke. If a stick or log is too long, you can burn one end in the fire. Gradually push it into the fire as it burns. Or you can burn it in half to get smaller pieces.

✓ At times the fire will begin to die out. To correct this, blow air (slow and steady) towards the bottom of the fire.

✓ Also, be sure that as sticks are added to the fire you leave some air space underneath each stick. Leaving air space ensures your fire will burn much better.

7

Fire Prevention

Smoke detectors are a true lifesaver — statistics have proven this.

Remove the following fire hazards from your home:

- Fireworks & oily rags

- Gasoline or flammable liquids inside the home.

- Flammable materials that are next to a furnace, heater, or fireplace.

- Some additional fire hazards are electrical outlets that are covered by furniture, clothing, window coverings, debris, etc.

Fire Prevention Checklist

☐ Have the local fire department come to your house and do a free Fire Prevention Inspection. This is probably the best way to find out about any specific fire hazards in your home.

☑ Install a smoke detector in each room and in each hallway. Test them once a month.

☑ Install an ABC rated fire extinguisher in the kitchen and on each level of your home.

☑ Have your fireplace or wood burning stove (and chimney) professionally cleaned ~~once a year~~.

☑ Keep space heaters at least three feet away from everything (walls, beds, clothes, etc.). Avoid using space heaters that use flammable liquids. Do not use an electrical space heater in wet or damp areas.

☑ Check all of the electrical cords in the house to make sure they are not cut or frayed. Do not overload electrical outlets by placing many plugs in one outlet. Make sure you have extension cords that can handle the power load of everything that is plugged into them.

☑ Remove any electrical cords underneath carpets or rugs.

☑ Make sure that electrical appliances (especially space heaters) are certified by Underwriters Laboratories (UL).

☑ Place a screen or glass insert on your fireplace to cut down on your fire risk.

☑ Don't leave lighted candles or lanterns unattended in your home. Keep matches and lighters away from children.

Plan an Escape Route

In the event of a fire, you want to have two escape routes out of your dwelling place. Plan these in advance and hold practice drills. Your local fire department can also show you how to set up escape routes.

8

Floods

Advance Preparation

Buy the Standard Supplies that are vital for any disaster. These are listed in the *SUPPLIES - ADVANCE PREPARATION* (page 68) chapter under the heading Standard Disaster Supplies. It is very important to read the entire chapter. Most of these items can be purchased from the company listed at the bottom of page 100. In addition to the *Regular Disaster Supplies*, you should stock the following *Special Supplies* in the event of a flood:

☑ Water pump   ☐ Boat or raft

☐ Life jackets   ☑ Sand bags

Preparation Tasks and Skills

☐ Learn in advance where your community evacuation routes are.

☐ Learn in advance the location of your local disaster shelters.

☐ Learn what the bomb/disaster warning sirens sound like.

☐ Read through this entire book so that you are more familiar with disasters.

☐ Purchase flood insurance. Check with your insurance agent for details.

Flood Watch

The flood watch tells you that your area may become flooded. If this is announced, do the following:

☑ Listen to the TV or radio for official instructions and information.

☑ Evacuate if you are told to do so by TV or radio.

☑ If there is time, and you believe that the flood will hit your street, you might want to move furniture, important papers, and valuables to a higher floor or to a safer location. Locate and unplug appliances.

☑ Fill up the car with gas if there is time. → Tank should already be full prior.

☑ Keep your 72-Hour Emergency Kit nearby.

☐ Gather the Standard Disaster Supplies listed in the SUPPLIES - ADVANCE PREPARATION chapter.

Flood Warning

A Flood Warning is very serious. This means that a flood is taking place now or is on its way. It is important to do the following:

☑ Listen to the TV or radio for official instructions and information.

☑ Evacuate if you are told to do so by the radio or TV.

☑ Turn off the gas, electricity and water if the officials have advised you to do so OR if you think the water will directly affect your home and cause flooding or structural damage. Read the chapter: TURNING OFF the GAS, ELECTRICITY and WATER (page 78).

☑ Take your 72-Hour Emergency Kit with you when you leave.

☐ Gather the Standard Disaster Supplies listed in the SUPPLIES - ADVANCE PREPARATION chapter (page 68).

☐ Be careful driving anywhere during a flood. If your car stalls or is disabled in the water, abandon your car immediately. Many people die when their car is swept away in flood waters.

☑ Don't drive through water unless you are positive how deep it is and that there is a solid surface underneath. For example, you see a bridge and it looks like there are only a few inches of water pouring over it. In reality, part of the bridge could be washed away by the flood. If you were to drive over it, your car would fall into the water. It is best to get out of your vehicle and investigate the full length and width of the bridge. Many times, however, the bridge could be severely weakened by the flood waters, making it still unsafe to cross even if the bridge appears to be intact.

After a Flood

☑ Help the sick and injured.

☑ Follow the official instructions given by radio or TV.

☑ Don't drink the faucet water until you read page 79.

☑ Contact your insurance agent regarding injuries or property damage.

9

Food

This chapter is about finding food in an emergency disaster situation. Read through this entire chapter before you gather and eat these foods. For information on long-term emergency food storage, read the chapter titled *SUPPLIES - ADVANCE PREPARATION* (page 68). People have lived for months without food. When you are looking for food, do not waste more energy looking for the food than you are going to get out of the food. **WARNING:** *Do not assume that because a plant, berry, or food is eaten by an animal it is safe for a human to eat. Birds, squirrels, and other animals eat foods that are poisonous to people.*

If you are dehydrated and low on water supplies, then you should eat very little food or none at all. The reason for this is that your body uses up water when it is digesting food. Eating when there is no water available will make you dehydrated. The exception to this would be if you are eating food that has a lot of water in it.

Vitamin C Deficiency (Scurvy)

WARNING: *Vitamin C deficiency, also called scurvy, is the first nutritional problem people would encounter within 4-6 weeks after their last vitamin C intake. The signs of scurvy are swollen or bleeding gums, followed by weakness, and bruises or wounds that will not heal. Death can result after long-term vitamin C deficiency. To prevent scurvy, eat seed sprouts (sprouting is explained below), chew on pine needle tips, or take a vitamin supplement.*

Seed Sprouting Instructions

WARNING: *Never eat pink coated seeds. They are very poisonous.*

Soak seeds for 12 hours in water. Drain completely. Place seeds in a quart size jar 3/4 inch deep or *less*. Then place a ball of wet paper inside the jar leaving a 1 inch air space. Do not cover with a lid. Keep in the dark. Rinse twice a day if possible. The seeds sprout in 2 to 4 days. Place in sun (if possible) just before eating to turn the leaves green. Eat the sprouts when they are 1/4 inch to 1/2 inch long for the best nutritional benefit.

Dangerous Foods

WARNING: *The following foods should not be eaten. Some of these are harmful and some are very poisonous.*

- Unknown plants
- Buttercup plant

14

- Mushrooms
- Unidentified berries (especially red or white berries)
- Any plants with milky sap or milky juice except dandelions.
- Seeds with dark or black mold, pink coating, or fungus (they can be poisonous).

  - Spiders
  - Centipedes
  - Millipedes
  - Caterpillars with hair
  - Gila Monsters

  - Moray eels
  - Barracudas
  - Raw snails
  - Newts
  - Toads

- Ocean Fish that do not have scales
- Raw Fish: Cook first because they have harmful parasites.
- Uncooked Birds: Cook first to kill parasites.
- Uncooked Insects, Worms, and Snails: These contain harmful parasites and must be cooked before eating.
- Rats and Mice: They can carry a variety of diseases including the newly discovered Hantavirus first reported in the four-corners area of the Southwestern United States.

Plant Safety Test

Never eat mushrooms, buttercups, or any plants with milky sap or juice (except dandelions.) Do not eat plants that you are not familiar with, unless it is absolutely necessary. If you are going to eat strange plants, first test them to see if you have any bad reaction to them such as:

- stomach ache
- diarrhea
- blurred vision

- vomiting or nausea
- numbness
- dizziness

- rash
- tingling
- headache

To Test an Unknown Plant:

☐ Place a very small amount (about the size of a pea) in your mouth. Keep the plant in your mouth for 15 minutes. If you have any burning, stomachache, or any ill effects, spit it out and do not use the plant.

☐ If there are no bad effects after 10 hours, take a teaspoon-sized bite of the plant and eat it.

☐ Wait 24-hours (one full day). If there are no bad effects, you can eat 1/3 cup of the plant.

☐ Wait an additional 24-hours (another full day). If there are no bad effects by this time, you can consider this an edible plant.

□ If at any stage of the test you experience any bad effects, then do not eat or test the plant any more.

Finding and Gathering Foods to Eat

The best way to gather food, and decide what type of food to gather, depends on where you are and what is available. Below are some of the many foods that can be eaten.

WARNING: *Never eat any plant that has a milky juice or sap (except dandelions.) Try to only eat plants which are known to be safe and that you can positively identify. NEVER assume that a plant (or other food) is safe because an animal eats it.*

Plants and Trees That Are Safe to Eat

I would strongly recommend you buy the book "Field Guide to Edible Wild Plants" by Bradford Angier.

- ACORNS: Remove the shells. Then to get rid of the bitterness (tannic acid), boil them until the water turns yellow. Pour off the water and add fresh water. Boil again and then rinse. Repeat if needed. You can dry the acorns and carry them with you.
- ALFALFA: Leaves and seeds are edible.
- CACTUS: The inside fleshy part of all cactus can be eaten.
- CATTAILS: Eat the roots (especially in winter), the young soft shoots, the white inside part of the stems and the yellow flower spikes. Eat raw or cook like corn. See the illustration page 45.
- CLOVER: The leaves and the stems can be eaten.
- DANDELION: The whole plant can be eaten.
- GRASS: Grasses are edible. One of the best parts to eat is the soft, white part that is located just below the ground. Make sure that what you are eating is grass. Do not eat if there is a milky juice or sap. Avoid grass with sawtoothed edges. **WARNING:** *Don't eat grasses if chemicals have been sprayed on them.*
- GRASS SEEDS: All grass seeds can be eaten. Grind into flour or soak overnight and eat. **WARNING:** *Never eat any grass seeds with a dark black mold, pink coating, or fungus on them. These are poisonous.*
- GREEN SEA PLANTS: This is the green, slimy stuff that you see growing on the bottom of boats and on the piers below the high tide mark. You can eat it raw or cook it. Gather from below the low tide mark. Don't eat if it came from very polluted waters
- INNER BARK OF TREES of aspen, birch, willow, most of the pines, and cottonwood trees, also known as black poplar: You can eat it raw or cook it like spaghetti. To gather, take off as large a piece of bark as you can get. Take lots of extra bark so you have something to eat later on. Peel off the inside of the bark — this is the light colored bark that is inside next to the outer bark of the tree. See the illustration on page 46. This may be some

of the only food that you can find in the winter. People who have been stranded in the mountains have been known to survive on this inner bark for months.

- **PALM TREES:** Eat the shoots of young palm trees, the inner white part of the stem (located just below the palm leaves), the coconuts, and the dates. The milk (liquid) inside coconuts is also very good to drink.

- **PINE NEEDLE TIPS:** Chew them to get vitamin C — very important in the diet, especially during the winter.

- **PLANTAIN:** The leaves are edible.

- **PURSLANE:** Eat the leaves and seeds. ***NOTE***: *The leaves contain lots of water. See the illustration on page 45.*

- **SEAWEED:** All types of seaweed can be eaten except the very thin kind that looks like thread. Eat raw or add some to soup. Make sure you gather seaweed from under the water level (under the low tide mark). Don't eat if it comes from very polluted water.

- **SUNFLOWERS:** Eat the seeds.

- **WILD MUSTARD:** Eat the leaves raw or cooked.

- **TREE SAP:** Maple, birch, aspen, hickory, cherry, and cottonwood (also known as black poplar) trees. Cut a small notch in the shape of a V in the trunk of the tree and let the sap drain out into a container. You should be able to collect up to a gallon of sap per day. Drink the sap raw or warm up and add other ingredients to make a soup.

INSECTS That Are Safe to Eat

All insects should be completely cooked, roasted, or boiled before eating. This is because they may contain harmful parasites. You can cook insects on a flat rock or on a stick next to a fire. Before you eat an insect, make sure that it is not on the list of dangerous foods (see pages 14 & 15). Here is a partial list of some of the insects that can be eaten:

- **ANTS:** Never eat ants that sting. Boil or roast whole before eating.

- **CATERPILLARS:** Do not eat them if they have hair or fuzz on them. Remove head and guts. Then boil or roast to cook before eating.

- **CICADAS, CRICKETS and GRASSHOPPERS:** First remove the head, wings, and legs. Then roast or boil. You can catch them by picking them up when they are cold, early in the morning just before sunrise. Another way to catch them is by laying out an all wool blanket at night. They are snared when their legs stick to the blanket.

- **HELLGRAMMITES:** In the eastern United States, these are found in streams underneath rocks. They look like flat, black/gray centipedes with pinchers on their heads. Remove the head and gut them. Cook, boil, or roast them before eating.

- **GRUBS and WORMS:** Remove head and guts. Then cook or boil.

ANIMALS and FISH That Are Safe to Eat

Boil, roast, or cook animals and fish before eating. If there are no utensils available to cook with, you can cook by placing the food on a flat rock or a stick next to a fire.

NOTE: Fish and animal bones can be crushed and cooked in soup. This is a good way to obtain calcium, which is a very important mineral for your health.

The following is a list of some of the more common animals that are available:

- ANIMALS can be attracted by using a salt lick. Use a block or pile of salt or make a salt lick by pouring very salty water on a rock & letting it dry, repeat 5-6 times. This is illegal. Do not do this unless it is an emergency.

- BIRDS: The easiest way to catch birds is with a baited fishhook on a line. To prepare: Gut them and remove head and feet. Next remove the feathers (this is much easier if you dip the bird in boiling water for 30 seconds first). Then cook completely (it will look like cooked chicken).

- FISH: The best way to catch fish (if you do not have regular fishing gear) is by spearing them or by making a simple fish trap. A fish trap can be made by sharpening thin sticks and pushing them into the bottom of a stream in a funnel pattern. You can also pile up stones in a funnel pattern if there are no sticks available. To make a funnel trap, see the illustration on page 46. When your trap is finished, go about 40 feet upstream and make lots of noise to scare the fish into the trap as you walk downstream towards the trap. To prepare for eating: Remove head and guts. Scale or remove skin, then bake, broil, or roast. *WARNING: Fish can have parasites and must be completely cooked before they are eaten. Cooked fish look whitish.*

- FROGS: The best way to catch frogs is to shine a light in their eyes at night. They will be unable to move. *CAUTION: This is illegal except in an emergency.* Another way to catch frogs is to make a long, double-pronged spear and spear them. To prepare a frog to eat: cut off the head and gut it. Remove the skin and then cook. *WARNING: Brightly colored frogs (red, yellow, blue, orange, etc.) are very poisonous. Also, do not eat toads.*

- LIZARDS: Cut off the head, legs and tail. Remove the skin and guts. Then cook. *WARNING: Gila Monsters are poisonous.*

- RABBITS: These are best caught with a Deadfall Trap or Snare placed in the path of their trail. See the illustrations on page 47. To prepare rabbits: Gut and skin them. You may boil, roast, grill, or bake them. *WARNING: The use of a Deadfall Trap or Snare may be illegal in your area, except in an emergency. Also, because rabbits can carry rabbit fever, only eat rabbits if there are no other sources of food available.*

- SNAKES: To prepare: Cut off the head and neck. Remove guts and then cook. Bury the head of a poisonous snake after you cut it off. *WARNING: If it is a poisonous snake, make sure it did not bite itself or you will be poisoned when you eat the snake.*

- SQUIRRELS: To catch squirrels, use a bag that can be made with a drawstring around it. Use a long pull string. Bait is placed inside the bag. After the squirrel enters, the bag is pulled closed with the long drawstring.

Squirrels can also be caught in Snares or with Deadfall Traps. See the illustrations on page 47. To prepare squirrels to eat: First gut them and remove the skin before cooking. Cook thoroughly: Boil, bake, roast, or grill.

- TURTLES: To prepare turtles: Cut off the head and legs. Remove the bottom shell and guts. Then cook. *NOTE*: *You can boil the turtle in its upper shell (the curved one). You may want to save the shell to cook in, store water, etc.*

10

Heat and Temperature — How to Get and Stay Warm

The following information should be very helpful to someone who needs to get warm or to stay warm. You may also want to read the chapters on *How to Build a FIRE* (page 9), *SHELTER* (page 63), *CLOTHING* (page 4), *SHOES* (page 66) and *WINTER STORMS* (page 83) for more information.

If Your House Has No Heat

In the *SHELTER* chapter you will find directions for building a quick shelter indoors that will keep you warm.

Do These Things to Get or Stay Warm

✓ Building a fire and then a shelter is very important when you are outdoors in a cold weather situation. Read the chapters on *CLOTHING, SHOES, How to Build a FIRE*, and *SHELTER* for more information. Also see the illustrations on pages 48 to 51 for shelter building.

✓ If you can't build a fire or shelter, keep moving and stay active.

✓ Take Vitamin C and Iron.

✓ Wear a hat, scarf and gloves. Over 50 percent of your body heat is lost through your head and neck. Your hands also lose a considerable amount of body heat.

✓ Wear three or four layers of dry clothing, towels, or 12 to 15 layers of newspaper wrapped around your arms and legs. Tie the towels or newspaper on with string. This will trap warm air next to your body.

✓ Drinking hot liquids is one of the best ways to quickly get your body warm.

✓ When it is extremely cold, keep your mouth covered with a cloth or a rag. Breathing in very cold air will quickly make your body cold.

✓ Lay close to (or hug) another person for warmth.

✓ Keep lots of insulating materials underneath you, especially when sleeping.

✓ Wear a hat when sleeping — fur or wool is best. Also wear socks. If your clothing becomes damp with perspiration, if possible, change to dry clothing before sleeping.

✓ Another very effective way of getting warm is to place very warm or slightly hot articles over your kidneys. This works well because your blood circulates through your kidneys about every two minutes. For example, you could heat up two canteens of liquid and place one over each kidney. The kidneys are located on the lower back just above the waist on the sides. See the illustration page 44. *WARNING: Don't do this to a person who is suffering from hypothermia (which is a dangerously low internal body temperature). You could cause severe heart problems or other complications. For more information, see the explanation of Hypothermia (page 22).*

Don't Do These Things — They Make You COLDER

√ If you are outside when it is cold, do not move around so much that you sweat. The sweat will freeze later and make you very cold.

√ Don't eat snow or drink very cold water. Melt the snow first and warm up any water before you drink it.

√ DO NOT get overly warm. Sweat (perspiration) chills you.

√ Do not let your hands and feet get cold. It is difficult to rewarm them once they become cold (because blood vessels constrict when they are cold).

√ Don't sit, sleep, or lay on anything that is cold such as the ground, rocks, metal, or snow. This quickly chills your body.

√ Don't smoke! It makes you colder because it constricts your blood vessels and you continually inhale cold air.

√ Don't drink alcohol! Alcohol is a depressant which slows the body functions and heart rate. Alcohol reduces feeling and sensation, thus giving you an illusion of warmth — this can be very dangerous! You think you feel warmer, but your body is actually colder. You are then at a higher risk of frostbite and hypothermia.

Converting Fahrenheit to Celsius

- Start with the temperature in Fahrenheit
- Subtract 32
- Multiply the answer by 5
- Divide the answer by 9 to get the temperature in Celsius
[(Temperature F - 32) x 5] ÷ 9 = Temperature in Celsius

Converting Celsius to Fahrenheit

- Start with the temperature in Celsius
- Multiply by 9
- Divide the answer by 5
- Add 32 to get the temperature in Fahrenheit
[(Temperature C x 9) ÷ 5] + 32 = Temperature in Fahrenheit

Signs and Symptoms of Frostbite

CAUTION: *A person may not have all of these symptoms, but may still be suffering from frostbite.*

Area of skin may look:

- white
- waxy
- flushed

- gray   • yellow   • bluish
- Affected area aches (dull ache), tingles, or is numb.
- Affected area no longer feels cold — possibly followed by a warm feeling.

Treatment for Frostbite

Get the frostbite victim to a doctor or hospital. Call 911 or your local emergency number if you are unable to transport the victim. If you are unable to get professional medical help for the frostbite victim, you may choose to begin the following first aid measures:

√ If the victim is alert, ask the victim if it is okay to treat him or her.

√ Heat some water to between 104° F and 109° F (40° C to 42° C) **WARNING: DO NOT use water that is hotter than 109° F.**

√ Place frostbitten area in the water until it looks red or flushed and is no longer cold to the touch. **WARNING: Do not leave the affected area in the water any longer than it takes to warm the area.**

√ If water is not available, place the frostbitten area under an armpit or on the abdomen (stomach area) of someone who doesn't have frostbite — otherwise, use your own armpit or abdomen to warm the area.

√ Place gauze between the fingers or toes if they are frostbitten.

√ Transport the frostbite victim to a nearby hospital as soon as possible.

√ DO NOT break any blisters that may develop — this opens the area to infection.

√ DO NOT rub, massage, or apply dry heat to the area — these actions may cause additional cell damage in the affected area.

WARNING: DO NOT thaw the area if you know it will most likely refreeze. This will cause more extensive cell damage than if the area had been left untreated. If there is no danger of refreezing, begin treatment for frostbite as listed above.

Hypothermia

Hypothermia is a condition in which the body can no longer heat itself. This condition is caused by below normal internal body temperature. A person is suffering from hypothermia when his or her internal body temperature reaches 95° F or lower. A person may suffer permanent heart damage, or their heart may stop when their internal temperature reaches 86° F. Death is imminent when the internal body temperature reaches 80° – 82° F or below. Hypothermia is a very serious and life threatening medical emergency.

The best way to avoid hypothermia is to dress appropriately in cold weather and build a fire and/or shelter. If this isn't possible, *keep moving.* If you are in an emergency situation in severely cold weather, and are not properly dressed, you can develop hypothermia in as little as 20 to 30 minutes. Remember that you are in danger of developing hypothermia in cold weather without a shelter/fire.

Symptoms of Hypothermia

CAUTION: A person may not have all of these symptoms, but may still be suffering from hypothermia.

- Reduced and/or shallow breathing
- Severe shivering
- Clumsiness, lack of coordination
- Pupils do not respond to light

- Skin is cold or pale
- Numbness
- Confusion
- Drowsiness

Treatment for Hypothermia

Hypothermia is a very serious medical condition. Call 911 if possible and get the victim to a hospital. If professional medical help is not available, you may choose to begin the following emergency first-aid measures:

- If the victim is alert, ask if it is okay to treat him or her.
- Take off any wet clothing. Dry the victim and dress in dry clothing.
- Wrap the victim in something with insulating value such as blankets, coats, or a sleeping bag.
- Lay down next to the victim and hold him or her close to help warm the person with your own body heat.
- DO NOT place the victim in hot or warm water. This can damage the heart or cause the heart to stop.
- DO NOT give the victim warm or hot liquids — for the same reason listed above.

Wind Chill Chart

Wind Speed ↓	Actual Temperature (°F)											
	50	40	30	20	10	0	-10	-20	-30	-40	-50	-60
	Equivalent Temperature (°F)											
Calm	50	40	30	20	10	0	-10	-20	-30	-40	-50	-60
5	48	37	27	16	6	-5	-15	-26	-36	-47	-57	-68
10	40	28	16	4	-9	-21	-33	-46	-58	-70	-83	-95
15	36	22	9	-5	-18	-36	-45	-58	-72	-85	-99	-112
20	32	18	4	-10	-25	-39	-53	-67	-82	-96	-110	-124
25	30	16	0	-15	-29	-44	-59	-74	-88	-104	-118	-133
30	28	13	-2	-18	-33	-48	-63	-79	-94	-109	-125	-140
35	27	11	-4	-20	-35	-49	-67	-82	-98	-113	-129	-145
40	26	10	-6	-21	-37	-53	-69	-85	-100	-116	-132	-148
Wind over 40 MPH does not have much additional effect	LITTLE DANGER of frostbite or hypothermia for correctly dressed person				HIGH DANGER of frostbite and hypothermia				EXTREME DANGER of frostbite and hypothermia			
	The young, elderly & sick become cold faster											

11

Home Security

What you are trying to do in securing your home is to make your home difficult and undesirable to break into. If a criminal does manage to break in, you should have small valuables, rare coins, jewelry and similar valuable items very well hidden. Your home should be set up to look like it is occupied at all times, even when no one is home. The following is a checklist that will help to increase your home's security:

☐ You can reduce crime in your area by up to 50% by starting or joining one of these groups. Write to:

NEIGHBORHOOD WATCH

LA Police Department, Devonshire Division

10250 Etiwanda Avenue, Northridge, CA 91325

NEIGHBORHOOD ANTI-CRIME CENTER

Citizen's Committee for NYC

305 Seventh Avenue, 15th Floor, New York City, NY 10001

☑ When you know you will not be home for an extended period of time and no other people will be in the house while you are gone, place a $10 or $20 bill inside the door, on the floor in a spot that is easily seen when you open the door. If the money is missing when you come home, you will know that someone has been there. Close the door, (the burglars/criminals may still be inside), leave and call the police.

☑ A large, barking dog can discourage burglars/criminals from targeting you.

☑ Don't show very expensive jewelry, flashy cars, or large amounts of cash in public. This may attract thieves.

☐ Get an unlisted phone number so criminals cannot use your phone number to find out your address in a real estate directory.

☐ Get a post office box and use it to receive your mail.

☐ If your state allows it, get a driver's license with a P.O. box as your address.

☑ Keep small valuables and important papers in a very well hidden, fireproof container or a small safe.

☐ Install a pick-proof lock on all entry doors. Call a locksmith for information.

☑ Keep all windows locked. Windows are the most common point of entry in a burglary.

☑ Sliding doors are frequently forced open. To help prevent this, place a wooden dowel or a long piece of broomstick at the bottom of the door in the door channel. Specially installed locks for sliding doors are also available at hardware stores or from your local locksmith.

☐ Walk around the exterior of the house and write down each place where you can see into the inside of the house. Cover each of these places with blinds or curtains so that criminals cannot see into the house and figure out if you are home.

☑ Remove any bushes that a burglar could hide in. *or trim them.*

☑ To light up your yard, install flood lights. The type that turns on automatically with an infrared or motion sensor are the best.

☐ Mark all valuables with state drivers license number, call local police for information.

☑ Place alarm warning stickers on ~~all~~ doors and windows. ~~Order from suppliers on the bottom of page 100 or at Radio Shack®.~~

☑ Get Caller ID, use it to see who is calling you and to block anonymous callers. (ask local phone company for details.)

☑ Install a peep hole in your entry door or mount a security camera near it (~~available from Radi~~o Shack®) and never open your door to anyone you don't know and trust.

☐ Have a motion detector alarm system installed in your home. Call ADT® 1-800-238-3009 or less sophisticated systems are available for under $100 from supplier at the bottom of page 100.

☐ Buy a couple of electronic timers that plug into the wall outlets. When you leave, plug the TV and a light into the timer. Set the timer to turn off the TV and light late at night or when the sun comes up.

☐ When you go out of town, have your mail and newspaper collected by a trusted neighbor instead of stopping delivery. Thieves may discover you are away when you have service stopped to your home.

☐ If you go out of town, ask a neighbor to park in your driveway so it looks as if someone is home.

☐ Leave out $20 cash and some "imitation" jewelry in a visible place in case of a burglary. The burglar will hopefully take these items and leave without looking further.

☑ Beware of dishonest workmen, handymen, repairmen and homeless willing to "work for food" (buy 'em a hamburger instead)— they have been known to come back and rob your house after seeing what is inside. Make sure any repairman is licensed, bonded and has no unresolved complaints at the Better Business Bureau.

12
Hurricanes

Read through this entire chapter for important information. Hurricanes are very serious and they have killed many people. If you are told to evacuate do so - your life may depend on it! Read the chapter on EVACUATION (page 8) for more details. **WARNING:** halfway through the storm there can be a calm period when the wind stops. This is called the "eye" of the storm. The hurricane winds will begin again 10 to 60 minutes later, with the wind coming from the opposite direction.

Hurricane Ratings

The strength of hurricanes are rated in the following manner:

Category		Wind speed (in miles per hour)
1	Week	74-95 MPH
2	Moderate	96-110 MPH
3	Strong	111-130 MPH
4	Very Strong	131-155 MPH
5	Devastating	155 + MPH

Advance Preparation

Buy the standard supplies that are vital for any disaster. These supplies are listed in the SUPPLIES - ADVANCE PREPARATION chapter (page 68) under heading Standard Disaster Supplies. It is very important that you read that entire chapter. Most of these supplies can be purchased from the company listed at the bottom of page 100. In addition to regular disaster supplies, you should stock the following special supplies in case of a hurricane.

Additional Special Supplies

[√] WEATHER INFORMATION RADIO: the type with an alarm
[ ] BOAT or RAFT
[ ] LIFE JACKETS
[√] ~~SILVER~~ DUCT TAPE: 3 **large** rolls
[ ] 2 x 4 BOARDS 15 to 40
[ ] PLYWOOD SHEETS: one for each window and door
[√] HAMMER, NAILS and CROWBAR

☑ FLOOD PUMP

☑ NYLON ROPE: use rope to tie things down *or paracord.*

Preparation Tasks and Skills

☐ Learn in advance where your community evacuation routes are.

☐ Learn in advance the location of your local disaster shelters. Practice driving to the nearest shelter once or twice a year.

☐ Learn what the bomb/disaster warning sirens sound like.

☐ Read through this entire book so you are more familiar with other types of emergencies and disasters.

Hurricane Watch

The hurricane watch is telling you that there is a threat of a hurricane hitting your area within 24 to 36 hours. Do the following things so that you will be prepared:

☑ Turn on the TV or radio and follow the official instructions. Evacuate if you are told to do so.

☑ Store water in bathtubs, buckets, or other containers. Store as much as you can — 30 gallons per person (or more) is good.

☑ Fill car fuel tank with gas. → *Tank should already be full prior.*

☑ Gather your Disaster Supplies. Read the Advance Preparation section above and the SUPPLIES - ADVANCE PREPARATION chapter (page 68) for more complete information.

☐ Read the next section for more information on how to make additional preparations.

Hurricane Warning

A hurricane warning means that a hurricane is going to hit your area within 24 hours. This is a very serious situation. Turn on the radio or TV and follow the official instructions. Review the following information to get prepared:

☑ Listen to the TV or radio for instructions.

☑ Gather needed disaster supplies and evacuate ~~to a shelter if you are told to do so~~ or if you feel that you will be safer.

If there is time, DO the FOLLOWING:

☑ Get your 72-Hour Emergency Kit and keep it with you. See the chapter on SUPPLIES - ADVANCE PREPARATION (page 68) for a complete listing of needed supplies.

☑ Gather 15 to 30 gallons (or more) of water per person.

☑ Fill the gas tank of your car.

☐ Board up the windows with plywood OR get duct tape and place large Xs across each window.

☐ Moor boats or move them to a safe area.

☑ Tie down or bring inside outdoor objects such as trash cans and bikes.

☑ Turn off the gas, electricity and water if the officials have advised you to do so OR if you think the hurricane will directly strike your home and cause flooding or structural damage. Read the chapter: TURNING OFF THE GAS, ELECTRICITY AND WATER.

☑ Move items to a higher floor or a safer place if you think that there will be flooding.

If You are Told to Remain in Your Home

☑ Stay tuned to your TV or radio for official information and instructions.

☑ Follow the preparation instructions listed above.

☑ Stay inside away from windows.

After the Hurricane

☑ Help the injured and the elderly.

☑ Listen to portable radio for instructions and information.

☑ Don't use any lighters, candles, matches, or any type of open flame until the gas lines have been checked for leaks.

☐ Do not use the phone unless it is a real emergency.

☑ Don't drink the water from the sink until the authorities on TV or radio say it is okay to drink. (The water may be contaminated.) Drink stored or bottled water. See page 79 for more information.

☑ Do not enter or stay in homes or buildings that have structural damage.

☑ Stay away from rivers, streams and coastal areas.

☑ If your home is flooded, have the electric company (or an electrician) turn off the electricity at the main circuit box. Also have the gas company (or a plumber) turn off the gas at the main shut-off valve. Then remove all water from the house.

☑ After the gas lines have been checked and the house is found to be safe, have the gas company turn on the gas.

☐ After the house and the appliances are completely dry, with the electricity turned off, plug in the electrical appliances. Then turn on the electricity at the main circuit box.

☑ Call your insurance agent if there are injuries or property damage.

13

How to Make Light

WARNING: *Never use candles, matches, or any type of flame after an earthquake or any disaster that may have damaged dwellings. This is because the gas lines may have been damaged and could be leaking gas.* **NOTE**: *A chemical light stick that you bend to activate is a very safe item to use for light after these type of disasters. These sticks will last at least 12 hours. See the bottom of page 100 for the name of a supplier.*

In a simple power failure when there is not a danger of a gas leak, you can make one of the following emergency lights:

Soda can lantern: Take a regular soda can and cut a square 2½ inches by 2½ inches out of the side and 1 inch from the bottom of the can. **NOTE**: *Peel back the metal instead of cutting it out. Then put an inch of sand or dirt in the bottom of the can. Place a small candle on top of the dirt or sand, twisting it into the sand to ensure that it will stand upright.*

Emergency oil lamp: You can use vegetable oil, melted fat, or engine oil for an emergency oil lamp. A wick can be made from string, a shoelace, yarn, or a thin strip of twisted cloth. Place the "oil" in a heat resistant container. See the illustration on page 44 for details. **WARNING:** *Do not use gasoline, kerosene, or anything that is highly flammable.*

14

Lightning

Lightning is very powerful static electricity. One of the leading causes of accidental death caused by nature is being struck by lightning — it can easily kill you if it strikes you. A lightning rod can be installed on your roof to help protect your home and its contents from a lightning strike. Another precaution that you can take to protect your valuable electrical appliances and equipment such as computers and stereos is to buy surge protector outlet strips to plug your expensive electronic items into. Available at hardware stores.

When Lightning Approaches

- To determine how far away the lightning is from your location: count the seconds between the time you see the flash of lighting and you then hear the thunder that follows. Divide the number of seconds by five. The answer will tell you approximately how many miles away the lightning is. (Example: You count 15 seconds between the lightning flash and the thunder. Divide 15 by five: the lightning is about 3 miles away.)

- Get inside a home, building, or car if possible.

- Keep away from telephones, electrical equipment, wires and metal heating radiators (or anything else that has an electrical current and/or contains metal). The electrical charge in lightning can enter the home through electrical wiring.

If You Are Outside

WARNING: *If your hair stands up, beware — lightning is about to strike! Quickly kneel down on the ground (don't lie down) and place your hands on your knees.*

√ Get inside a home, building or car (if possible).

√ Do not stand under a tree. Stay away from any water, trees, power lines, or any object that contains metal.

√ Stay in a low area such as a depression in the ground or a cave until the storm leaves.

√ Do not go into a dry creek bed because of the danger of a flash flood which might follow a thunderstorm.

15

Lost or Stranded

People who survive becoming lost or stranded usually return alive because they didn't panic. They carefully thought out a plan or strategy for rescue or return. Remember — panic and serious mistakes in judgement associated with panic are the biggest killers of lost and stranded people. Your brain and a good attitude are your main survival tools. This cannot be emphasized enough.

- If you have a CB radio installed in your vehicle, or carry a portable CB radio, you can use Channel 9 to call for help. The police regularly monitor Channel 9.

- If you are able to get to a pay telephone and don't have any coins, you can usually still dial a friend or relative for help by using the following toll free numbers:

    1-800-OPERATOR   or   1-800-COLLECT
    (1-800-673-7286)   or   (1-800-265-5328)

- To have money wired, call 1-800-CALL-CASH

Stranded

If you are stranded, look over the following information for help:

✓ Try praying. Many lost people have reported miracles happening through prayer.

✓ Stop, rest, be calm and think out a plan. Read the *PRIORITIES* chapter (page 59).

- Make signals for rescue. See the *SIGNALS* (page 67) chapter for instructions.

✓ Drink lots of water — at least a half gallon a day. See the chapter on *WATER* (page 79) for more details and information.

✓ If you decide to stay where you are, start building a shelter and fire five to six hours before sunset. Read through this book, especially the *SHELTER* (page 63) and *How to Build a FIRE* (page 9) chapters for ideas and information.

- If you are not dressed appropriately for the existing weather conditions, see the chapters on *CLOTHING* (page 4) and *SHOES* (page 66) for improvising emergency clothes.

- See the chapter on *HEAT and TEMPERATURE: How to Get or Stay Warm* (page 20) for information on staying warm in cold weather and for information on hypothermia and frostbite.

- If you are stranded during a winter storm, see the chapter entitled *WINTER STORMS* (page 83) for more information.

✓ If you are stranded by the roadside or your car is broken down, you can signal to a passing trucker to call for help. Use one hand to point at the truck and get the truck driver's attention. Use your other hand to make the form of a telephone receiver and mouth the words, "Help." Most truckers can radio to the nearby town by CB radio (Channel 9) to have help sent to your location. See *Illustration* on page 55.

✓ In most cases, you should stay where you are and wait for rescue. For example, it is usually better to stay near your car and the road than it is to head out into unknown territory. However, sometimes it *is* better to walk out. Deciding whether to stay or walk out depends on the situation.

Deciding to Walk Out

CAUTION: It is very common for people who are lost to walk in circles. To prevent this, choose the most distant object or landmark in the direction that you would like to travel. Then walk a straight path towards the object to prevent walking in circles. When you reach that point, choose another object or landmark in the distance and repeat the same process.

☑ Be certain that leaving may help and won't create further problems such as lack of food or water.

☑ Be certain that you can bring or acquire water and food while traveling to safety.

☑ Have a general idea of where you are going and how far it is; determine that you are capable of traveling that far.

☑ Know that you can maintain a direction of travel by using a compass or other means. Read the DIRECTION FINDING chapter (page 5).

☑ Be confident that your traveling companions can be trusted, for example, not to panic or steal the food and will not cause serious problems that could make the situation worse.

☑ If you do walk out, leave a note for potential rescuers. Also remember to leave notes or signs along the way for rescuers to follow. (Read the SIGNALS chapter (page 67) for more information). Important information to leave for a rescue or search party includes:

- your direction of travel
- date and time you left
- needed supplies such as water
- how many people there are
- any problems or injuries
- your names; people to notify

On page 88 of this book there is a form you can fill out (or duplicate) and leave for a rescue party or the police.

16

Mental Help in an Emergency

Disasters and emergencies can cause people to experience a lot of mental or emotional pain and problems. The following guidelines work well for people who are having mental or emotional problems as well as those who are suffering from a loss.

√ The first step in an emergency is to prevent panic and to calm a very upset person. Show that you are a friend and want to help.

√ Let the person know that you care about him or her and the problems the individual is suffering. Give reassurance that you will listen to concerns and grief without judging.

√ Have the person talk about whatever he or she wants to discuss. Don't criticize or argue even if the individual seems to be hostile. Let the person talk as long as needed.

√ Allow the person to cry and to grieve about the problems, lost or injured people, or lost possessions. *NOTE: This is a very important step in the recovery process. It is also very important for people to express their feelings. They need to have someone there who is a good listener and really cares about them and their situation.*

√ If the situation is very serious, refer the person to a professional counselor.

• If you are experiencing mental distress, be sure to seek out help.

17

Nuclear Accident

Advance Preparation

The *Standard Supplies* that are vital for any disaster are listed in the chapter titled SUPPLIES - *ADVANCE PREPARATION* (page 68) under the heading *Standard Disaster Supplies*. It is very important to read the entire chapter. In addition to those supplies, you should stock the following *Special Supplies* in case of a nuclear accident. Most of these items can be bought from the company listed at the bottom of page 100.

- ☑ POTASSIUM IODIDE PILLS: 1 bottle per person. (These pills help to protect the thyroid in case of NUCLEAR ACCIDENT or WAR.)

- ☐ PLYWOOD: 8 sheets (4 ft. x 8 ft. size) for building shelter

- ☐ SANDBAGS: 50 to 100

- ☑ SHOVEL: 1 for each person

- ☐ 2x4 AND 4x4 BOARDS: 15 of each size

- ☑ FIRST AID BOOK with a section on radiation exposure.

- ☑ COMPREHENSIVE SURVIVAL BOOK with a section on Nuclear War and building Nuclear Shelters. ~~Nuclear War Survival Skills by Cresson Kearny is an excellent book~~.

Preparation Tasks

- ☐ Learn where your community evacuation routes are.

- ☐ Learn the locations of the local disaster shelters. Practice driving to them once or twice a year.

- ☐ Learn what the bomb/disaster warning sirens sound like.

- ☐ Build a bomb/radiation shelter in your yard or basement.

- ☐ Read through this entire book for information on preparing for any type of emergency situation you might encounter.

Radiation

The radiation from a nuclear power plant accident could range from harmless to deadly. In the event that radiation leaks outside a nuclear plant, get inside a radiation shelter immediately. Keep a TV or radio on so you can receive official information and instructions. Be sure to follow any instructions that are broadcast. In the event of a radiation leak outside the plant it is very important that everyone take potassium iodide pills to help protect the thyroid gland.

Radiation Danger Levels in a Nuclear Accident

Radio and television broadcasters will air reports during a nuclear accident. The following types of broadcast messages represent different levels of radiation danger:

- **NOTIFICATION OF AN UNUSUAL EVENT:** This means that there is a problem at a nearby nuclear power plant, no leak is expected at this time. Stay tuned to the radio or TV for more information.

- **ALERT:** This means radiation might leak, or is leaking. It is not expected to leak outside the plant at this time.

- **SITE AREA EMERGENCY:** Some radiation could or is leaking into the outside air. Follow instructions of local officials. Take your potassium iodide pills, if advised by officials, to help protect your thyroid gland (be sure to follow instructions on the bottle).

- **GENERAL EMERGENCY:** THIS CAN BE VERY SERIOUS. A LARGE AMOUNT OF RADIATION MAY BE LEAKING OUTSIDE THE PLANT. Take your potassium iodide pills, if advised by officials, to help protect your thyroid gland (be sure to follow instructions on the bottle).

If a state of GENERAL EMERGENCY is declared you should:

√ Follow the instructions broadcast on the TV or radio.

√ Take potassium iodide pills, if advised, to help protect your thyroid gland.

√ Protect yourself from the radiation by getting into a shelter if possible.

√ Evacuate the area if you are told to do so on the radio or TV. If you evacuate in your car, keep your mouth and nose covered with a thick cloth or a rag. Keep the car sealed off from the outside air by turning off the heater, vents and air conditioner, and rolling up the windows.

If radiation is leaking into the air and you are TOLD TO STAY HOME, you should:

√ Take potassium iodide protection pills to protect your thyroid gland.

√ Seal off your house from outside air in the following ways: close all windows, doors and any opening to the outside air. Turn off the air conditioner, heater, fan and vents. Stay in the basement or the lower part of a house. Surround yourself with as much dense materials on all sides to protect yourself from the radiation. You can use almost any dense materials such as earth, sandbags, books, canned food. Read the chapter titled nuclear war for more instructions on building a shelter.

√ Don't go outside unless it is vital. If you do go outside, only stay a very short amount of time. Cover your mouth and nose with a thick cloth or a rag. When you come back inside, remove all of your clothes and shoes, then throw all of your clothes and shoes outside. Take a very thorough 10 minute shower. Put on fresh clothes and shoes that were not exposed to the radiation.

18

Nuclear War

Advance Preparation

Buy the Standard Supplies that are vital for any disaster. These are listed in the SUPPLIES - ADVANCE PREPARATION chapter (page 68) under the heading Standard Disaster Supplies. It is important to read the entire chapter. Most of these items can be purchased from the company listed at the bottom of page 100.

In addition to the Regular Disaster Supplies, you should stock the following Special Supplies in case of a Nuclear War:

☑ RADIATION METER or badge

☑ POTASSIUM IODIDE PILLS: 1 bottle per person (These pills protect the thyroid gland in case of NUCLEAR WAR or a NUCLEAR ACCIDENT.)

☑ SHOVELS: 1 for each person

☐ SAND BAGS: 50 to 100

☐ PLYWOOD: 10 sheets (4 ft. x 8 ft. size)

☐ 2 x 4 and 4 x 4 Boards: 15 of each size

☑ WOOL BLANKETS and warm clothing

☑ FIRST AID BOOK with a section on radiation exposure

☑ COMPREHENSIVE SURVIVAL BOOK with a section on Nuclear War and building radiation shelters. ~~Nuclear War Survival Skills by Cresson Kearny is an excellent book~~.

☑ PORTABLE RADIO: Preferably the hand crank type that doesn't need batteries. See bottom of page 100 for the name of a supplier. *NOTE: Store the radio wrapped in several layers of aluminum foil to protect it from electrical damage that can occur during a nuclear attack.*

Preparation Tasks and Skills

☐ Learn in advance where your community evacuation routes are.

☐ Learn in advance the location of your local disaster shelters. Practice driving to them once or twice a year.

☐ Learn what the bomb/disaster warning sirens sound like.

☐ Read through this entire book so that you are more familiar with what to do in all types of disasters.

☐ Always store your 72-Hour Emergency Kit and other emergency supplies in a place that is easy to get to.

☐ Build a bomb shelter in your home or yard or designate a place in your home or yard to build a bomb shelter. If you build a shelter in your home, a basement location is the best choice.

Warnings and Other Information

In a nuclear attack, the sky will become very bright — more so than you have ever seen before. You will hear far away explosions and there may be a sudden power outage. If you see these things happen, take cover immediately (hopefully in a shelter). *Warning: DO NOT look at the explosions. They can blind you from as far away as 100 miles.*

The sirens should go off if nuclear bombs are on the way. Depending on the method used to launch the bomb(s), you will only have from three to 40 minutes before they land. *NOTE: When you see things getting really serious, go to (or build) a bomb shelter. The time to build a bomb shelter is before the bombs are fired. It will take about 48 hours to build a good shelter.*

If bombs are on the way or an attack seems very likely, you should take your potassium iodide pills to help protect your thyroid gland. Be sure to follow the instructions on the bottle.

NOTE: One tactic in a nuclear war is to explode nuclear bombs at a very high altitude — 200 miles above ground. If this happens, it would permanently destroy almost anything with electronics: all TVs, telephones, car electrical systems, computers and almost all other electronic equipment.

The best protection from radiation is to shield yourself in a bomb/radiation shelter as described below. There is a bomb shelter space for almost every person in Switzerland, Russia, and the big cities in China. I believe that we should have the same protection for our citizens.

WARNING: If a nuclear bomb is exploded, do not come out of a shelter for up to two weeks after the last bomb is exploded — or until the officials on the radio tell you it is safe. This is because the radiation falling from the sky could kill you in as little as one hour.

Protection Methods

The following are the main ways in which you can protect yourself from nuclear fallout effects. It is recommended that you consult books on Nuclear War Protection for more details.

Shielding

You can reduce the amount of radiation up to 1/1000th of the amount that you would receive outdoors by building a temporary or a permanent radiation/bomb shelter. To achieve this amount of protection you will need to be surrounded by three feet of packed earth or the equivalent in protection factors — 22 inches of concrete or 88 inches of wood. You can build a shelter in a basement or dig a shelter in the ground outside.

You must have an air intake vent which should have a filter on the end that is outdoors. A filter is not essential but an air intake vent is vital — especially in warm or hot weather. You can make a filter by cutting a circular piece out of a house furnace filter. The air vent tube should be 5 to 10 feet long. *Warning: Temporarily block air intake vent if sand-like particles are falling (these particles contain a lot of radiation). A comprehensive survival book will give detailed instructions on building shelters. The book entitled The Sense of Survival by Alan South or Nuclear War Survival Skills by Cresson Kearny are good choices.*

Distance

Be as far away from the place where the bomb explodes as possible (50 miles or more is a good distance). The farther away that you are, the less radiation you receive. This is not always practical, because you might be trying to escape when a bomb strikes — in which case you would have little or no protection. One of the main ways in which people are hurt from nuclear weapons is when the blast or pressure wave hits them. Also, there may not be time enough to escape or escape routes might be too crowded or dangerous.

Time

Time is a very important factor. You will need to stay inside a shelter for up to two weeks after the last bomb explodes. Bombs could be dropped for weeks after the first one strikes.

WARNING: Leaving your shelter while radiation is still falling could kill you in as little as one hour. DO NOT come out of your shelter until the authorities tell you by radio or TV that it is safe to come out.

19

Power Failure and Blackout

Advance Preparation

Buy the Standard Supplies that are vital for any disaster. These are listed in the SUPPLIES - ADVANCE PREPARATION chapter (page 68) under the heading Standard Disaster Supplies. It is very important to read the entire chapter. Most of these items can be purchased from the company listed at the bottom of page 100. In addition to the Regular Disaster Supplies, you should stock the following Special Supplies in case of a power failure:

☑ Portable stove and fuel such as wood burning or propane.

☑ Portable grill (or hibachi) and charcoal to use outdoors for cooking.

Long Term Power Failure Choices in Cold Weather

If the power fails and it is cold, your choices are:

- Winterize (protect) the home from freezing and leave for another location which still has power (see section below).

- Winterize your home to protect from freeze damage and continue to live in your home. You would need to use an alternate emergency heat source such as a wood or propane stove in your living/sleeping area (see section on following page).

WINTERIZE and GO

The reason for winterizing your home is to protect your pipes and anything that holds water from freezing. There are two ways of doing this. The best method is to shut off the water at the street valve. Read the chapter TURNING OFF the GAS, ELECTRICITY and WATER (page 78) for detailed instructions.

If you cannot shut off the water at the street, then shut off the main water valve inside the house. This is a round faucet handle which is located where the water enters into your house. Turn the handle to the right (clockwise) to shut the water off. See the illustration on page 56.

After you have shut off the water supply, find the lowest sink or faucet in the house and turn it on so that all of the water will drain out. Don't turn it off until the water has completely stopped running. Then flush all of the toilets so that there is no water left in them. If you cannot get all of the water out of your toilets, pour some antifreeze (recreational vehicle type is best) into the tank and the bowl of the toilet (top and bottom) to protect them from freezing, then breaking. Then drain the water heater: first turn off the gas and water supply to the tank. Next, turn on the small valve or faucet at the bottom of the water heater. *NOTE: You may need to open the valve on the top of the water heater before the water will flow out.*

STAY and USE ALTERNATIVE HEATING

NOTE: See Shelter Section for how to build a quick shelter. A propane, wood stove, or an oil burner are some of the safer heating units. If you have to use a propane or kerosene heater, open a nearby window so that you have plenty of ventilation. Do not use a charcoal grill indoors: the fumes can kill you.

It is much easier to heat just one or two rooms, or one level of the building or home, than it is to heat the entire house. If you only use a few rooms or only one level of the home, pour some antifreeze into the toilets (tank and bowl) in the unheated areas to protect them from freezing, and breaking. Read the section above for details.

20

General Preparation Tasks and Skills

It is very important to learn and prepare the following things before a disaster or emergency occurs. This can make the difference between life and death when the disaster strikes. Practice your preparation skills in advance so that you will be better prepared when an emergency arises or a disaster strikes.

☑ Keep a disaster instruction book such as this in an easy to find place.

☐ Read through this entire book so that you know what is needed in the case of different types of disasters.

☐ Build a disaster shelter in your home or yard (underground). You can call your local government for free information and plans.

☑ Have all of your preparedness supplies bought and stored in an easy to find place. Read the chapter titled SUPPLIES - ADVANCE PREPARATION (page 68) for details. Remember to also store supplies in each car and at your work place.

☐ Know the location of the nearest disaster shelter.

☐ Practice driving to the nearest disaster shelter at least once or twice a year.

☐ Learn what the evacuation routes are in case of disaster in your community.

☑ Locate in advance the radio station with the strongest signal that will be broadcasting emergency information. Write the station frequency number on your portable radio.

☑ Select a person who lives out of state who you can call in a disaster. This person can relay messages to friends and relatives and help to reunite people.

☑ Pick out a place in advance where friends and family can reunite if they are separated during an emergency. A relative, friends house or shelter is a good location to meet people. Have an alternate location to meet in the event that you can't reach the first location or encounter unforeseen problems.

☑ Keep your 72-Hour Emergency Kit in an easy to find and accessible place.

☑ Learn how to light a fire with a match and wood. See the chapter on How to Build a FIRE for details (page 9).

☑ Know the location of your gas, water and electric utilities. Record their location on page 89. Learn how to turn them off. See the chapter on TURNING OFF the GAS, ELECTRICITY and WATER for more details (page 78). Also see the illustrations on pages 56 to 58.

☑ Learn first aid and CPR. The Red Cross and local hospitals teach classes.

☑ Learn how to operate an emergency flood pump.

☑ Learn how to build a temporary shelter. See the SHELTER (page 63) chapter for help.

☑ Learn how to sharpen a knife and axe with a ~~diamond type~~ sharpener.

☐ Purchase earthquake and flood insurance. Ask your agent about replacement value versus prorated value.

☐ Have your insurance agent explain what forced entry and mysterious disappearance are. You may want to add these to your insurance policy.

☐ Take a complete household inventory of all possessions (for insurance purposes). Write them down and take pictures or videotape the items. Store your records in a safe deposit box.

☑ Join an auto club such as AAA®. These organizations offer towing and repair services. They also will help you map out the safest routes to travel and help you arrange for lodging. They also offer free travelers checks.

☐ Consider adding towing coverage to your car insurance policy. Some insurance companies will reimburse your towing expenses for an additional premium of $5 per year.

☐ Carry light sticks in the trunk of your car to use along the roadside when you are stranded or need help.

☑ Always carry water and a complete 72-hour kit in your car for emergencies.

Practice Disaster Drill

☐ Conduct a practice Disaster Skills Drill in your home when the weather is good. Temporarily shut off all of the utilities: water, electricity, and gas. Stay inside your home for one to three days and try to live off the supplies and equipment you have stored for an emergency. You will quickly learn what skills, equipment and supplies you would need in a real emergency. Have the gas company turn on your gas after your drill. *NOTE*: *Inquire about any fees or charges for this service before having the gas or water turned off for a drill.*

For the ultimate test: Try surviving in your backyard in the middle of winter using only the supplies available in your 72 hour kit, you will quickly find out the different supplies and skills that are needed in a worst case scenario.

Fire and Warmth

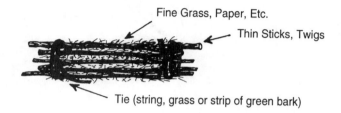

Fine Grass, Paper, Etc.

Thin Sticks, Twigs

Tie (string, grass or strip of green bark)

Tinder Bundle

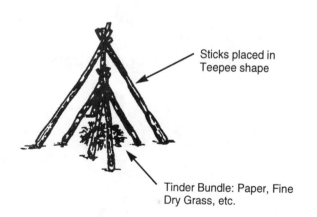

Sticks placed in Teepee shape

Tinder Bundle: Paper, Fine Dry Grass, etc.

Fire Setup

Fire and Warmth

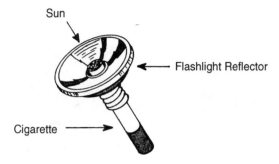

Sun

Flashlight Reflector

Cigarette

Improvised Fire Starter

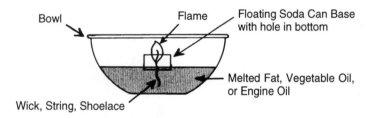

Bowl

Flame

Floating Soda Can Base
with hole in bottom

Melted Fat, Vegetable Oil,
or Engine Oil

Wick, String, Shoelace

Emergency Light

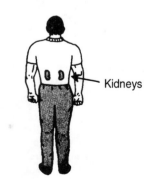

Kidneys

Location of Kidneys

Food

Cattail Plant

Purslane Plant

Food

Tree Trunk

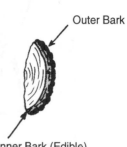

Outer Bark

Inner Bark (Edible)

Edible Tree Bark

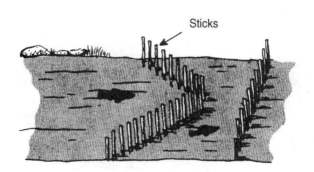

Sticks

Funnel Trap

Food

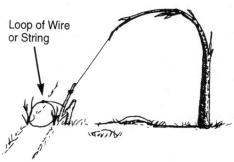

Loop of Wire
or String

Bent Over Sapling

Snare

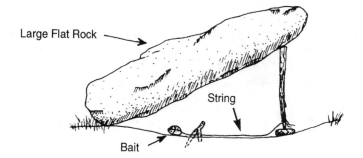

Large Flat Rock

String

Bait

Deadfall

Shelter

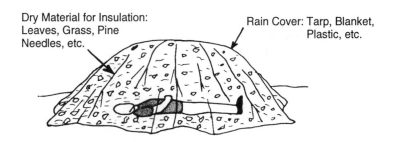

Dry Material for Insulation: Leaves, Grass, Pine Needles, etc.

Rain Cover: Tarp, Blanket, Plastic, etc.

Squirrels Nest Shelter

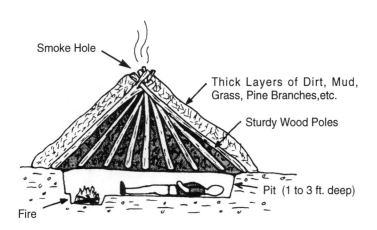

Smoke Hole

Thick Layers of Dirt, Mud, Grass, Pine Branches, etc.

Sturdy Wood Poles

Pit (1 to 3 ft. deep)

Fire

Semi-Permanent Shelter

Shelter

Cover with a Plastic
Tarp, Branches or
Branches and a Blanket.

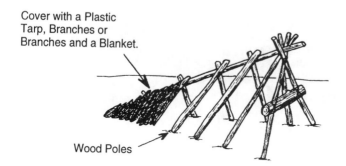

Wood Poles

Pole Shelter

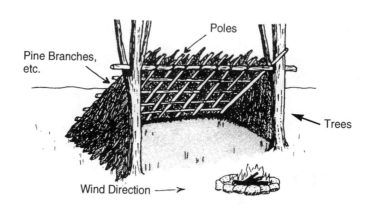

Poles

Pine Branches,
etc.

Trees

Wind Direction ⟶

Lean-to Shelter

Shelter

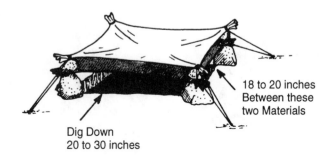

18 to 20 inches
Between these
two Materials

Dig Down
20 to 30 inches

Desert Shelter

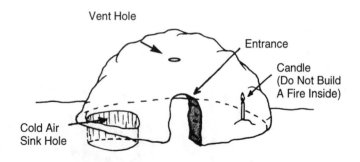

Vent Hole

Entrance

Candle
(Do Not Build
A Fire Inside)

Cold Air
Sink Hole

Snow Shelter

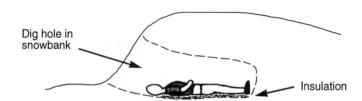

Dig hole in
snowbank

Insulation

Quick Snowbank Shelter

Place 5 to 15 inches of insulation underneath you (read page 65 for details).

Shelter

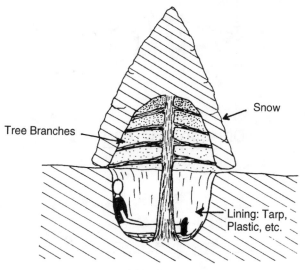

Snow

Tree Branches

Lining: Tarp, Plastic, etc.

Snow-Tree Shelter

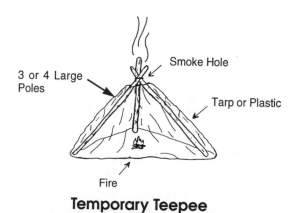

Smoke Hole

3 or 4 Large Poles

Tarp or Plastic

Fire

Temporary Teepee

Snowshoe

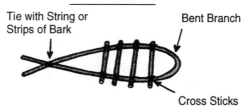

Tie with String or
Strips of Bark

Bent Branch

Cross Sticks

Direction Finding

Approximate
North

Setting Sun

Point as shown. Right hand
toward rising sun, left hand
toward setting sun.

Rising Sun

South

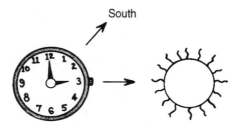

Point Hour Hand at Sun.
South is half way between hour hand and 12:00
If the watch is on Daylight Savings Time subtract one hour from
watch time first. (Example: It is 4 pm Daylight Savings Time subtract
one hour. South is half way between 12 and 3 pm. Not 12 and 4 pm.)

Direction Finding

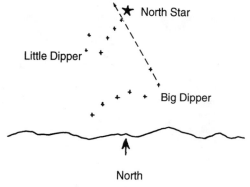

North

The North Star is within
1% of True North

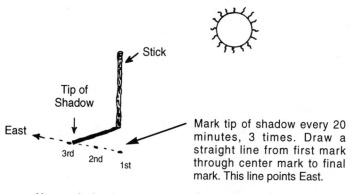

Mark tip of shadow every 20 minutes, 3 times. Draw a straight line from first mark through center mark to final mark. This line points East.

Use sun in the daytime and the moon at night.

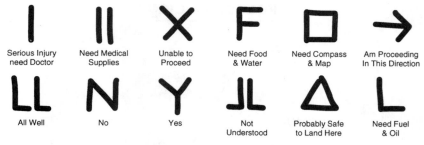

I	II	X	F	□	→
Serious Injury need Doctor	Need Medical Supplies	Unable to Proceed	Need Food & Water	Need Compass & Map	Am Proceeding In This Direction

LL	N	Y	⅃L	△	L
All Well	No	Yes	Not Understood	Probably Safe to Land Here	Need Fuel & Oil

These signals should be at least 25 feet long.
Airplane will rock wings side to side if signal is understood.

Ground to Air Signals

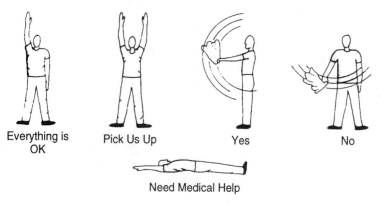

Everything is OK	Pick Us Up	Yes	No

Need Medical Help

Airplane will rock wings from side to side if signal is understood.

Ground to Air or Person to Person Signals

A •–	H ••••	O – – –	V •••–
B –•••	I ••	P •––•	W •––
C –•–•	J •–––	Q ––•–	X –••–
D –••	K –•–	R •–•	Y –•––
E •	L •–••	S •••	Z ––••
F ••–•	M ––	T –	SOS •••–––•••
G ––•	N –•	U ••–	

Use flashlight, signal mirror, etc. Dash (–) is two seconds. Dot (•) is a half second.

International Morse Code

Signals

Turn on the car's emergency flashers.
With one hand, make the shape of a phone,
and with the other, point at the driver.
Then mouth the words "Help!"

Signaling Trucker for Help

Water Shut-Off

Shut-Off Valve

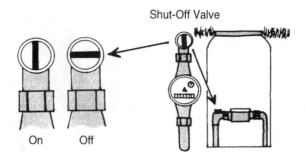

On   Off

Outdoor Street Water Valve

Note: *Valve may only turn in one direction* (see page 78)

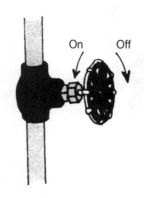

On   Off

Indoor Water Valve

Electricity Shut-Off

Main Shut-Off

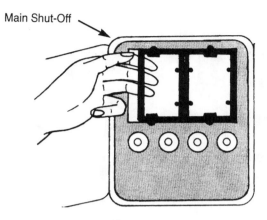

Fuse Box

Warning: *Read shut-off instructions on page 78 before beginning*

Main Shut-Off

Individual Circuit
Breakers

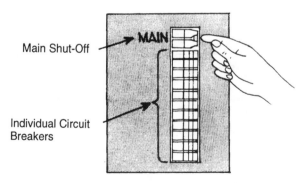

Circuit Breaker Box

Warning: *Read shut-off instructions on page 78 before beginning*

Gas Shut-Off

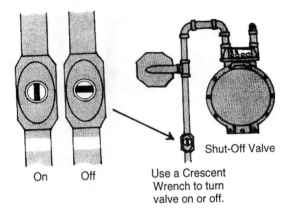

On   Off

Shut-Off Valve

Use a Crescent
Wrench to turn
valve on or off.

Outside Gas Meter/Shut-Off

Caution: Read page 78 before attempting to shut off gas.

21

Priorities

Remember that your brain is your best survival tool. Regardless of the disaster or emergency that you find yourself in, you should first figure out what is the greatest immediate hazard. Work to fix that problem, then work on your next greatest threat. For example: If it is very cold, say -20°F, your greatest danger is the cold. So you first work on building a fire and next a shelter. Then you go on to your next priority such as water and food. Your order of priorities depends on where you are, what you have available around you and many other factors. You must decide what your priorities are. The following may be some of the priorities that you may want to consider:

- FIRE
- FIRST AID
- FOOD
- WATER
- RESCUE
- SHOES
- SHELTER
- SIGNALS
- CLOTHING
- TURNING OFF UTILITIES (Gas, Electricity, Water)
- PROTECTION FROM HARMFUL ANIMALS OR PERSONS

22

Riots

Advance Preparation

Buy the Standard Supplies that are vital for any disaster. These are listed in the SUPPLIES - ADVANCE PREPARATION (page 68) chapter under the heading Standard Disaster Supplies. Most of these items can be bought from the company listed at the bottom of page 100.

Preparation Tasks and Skills

☐ Learn in advance where your community evacuation routes are.

☐ Learn in advance the location of your local disaster and emergency shelters.

☐ Learn what the bomb/disaster warning sirens sound like.

☐ I would highly recommend that everyone buy the book called The Truth About Self-Protection, Bantam Books®. This book will teach you how to protect yourself in a variety of situations. During a riot, the crowds will do things that individual people would not normally do.

☑ Most people who are killed in riots are beaten to death by rioters or are trapped in fires. Prepare an escape route in advance so that if the place that you are in catches on fire, you can get away. This escape route can also be used if the rioters are trying to break into where you are staying.

☑ Try to leave the area before a riot gets started. If you are caught out in a riot, try to get into the nearest safe place such as hospitals, large hotels, areas with lots of police, ~~or official emergency shelters~~.

During a Riot

WARNING: DO NOT STOP YOUR CAR IN A RIOT AREA FOR ANY REASON! If you stop, you may be pulled out and beaten or killed. Keep driving! When you come to intersections, if you are in any danger, make a right turn instead of stopping.

☑ If you are inside your home and rioting is taking place outside, it is usually best to stay inside. Don't let the rioters hear or see you. If you are listening to the news of the riot on your radio or TV, keep the volume very low so people don't know that you are home. You have to be seen or heard for people to know where you are — and if they don't find you, it is hard for them to hurt you.

☑ Don't go sightseeing to a riot area. If you are caught in a riot, keep driving until you are far away from the riot. If your car will not run, try to get into a building with plenty of people inside it. Read the section above for details.

23

Sanitation and Waste Disposal

After disasters, very harmful germs such as typhoid and cholera are spread by poor sanitation and improper waste disposal. To help prevent the spread of disease, read the following information:

✓ Always wash your hands before eating or handling food and especially after going to the bathroom or handling anything that contains germs. Also, pay extra attention to hand washing when you are ill, have an infection or open wound, or are caring for anyone who has illness, infections, or wounds.

✓ Disinfect toilet seats before using with a bleach solution (see below; same as used for dishes)

✓ Wash clothes often, especially socks & underwear

✓ You can take a sponge bath with 2 cups of water by doing the following: Take an 8 inch square of cloth and stand on something to keep the floor dry such as newspaper or a towel. First dip cloth in 1 cup of warm water that has a very small amount of soap in it. Wash face first (to wash hair requires much more water) then work down over entire body to the feet. When done rinse off with a full cup of water, by dipping two fingers in the water & rubbing skin from head down to the feet. Boil the cloth if you are going to reuse it.

✓ Don't let flies or other insects land on food or walk on food preparation and eating surfaces.

• Don't let animals including domestic dogs and cats lick your hands, face, or elsewhere on your body.

✓ Take measures to keep disease-carrying mice, rats, or other rodents and animals away from food sources and waste disposal areas.

✓ Dishes, knives and forks can be sterilized by soaking them for 30 minutes in a gallon of water that has two tablespoons of plain liquid chlorine bleach added to it. REMEMBER: Don't use powdered bleach because it is poisonous. Also, don't use color safe or scented bleach because they may contain harmful chemicals.

Waste Disposal

✓ If a working toilet is not available, all waste must be treated with chemicals or burned to kill diseases and germs. After treatment with chemicals or burning, the waste should be held for future disposal in a working toilet or buried.

✓ Keep rodents and flies away from waste.

✓ If you are using a latrine, pour some treatment chemicals over the waste after each use and then place some dirt on top. To treat the waste, you can purchase prepared sewage treatment chemicals or you can make your own chemicals by adding five tablespoons of plain liquid chlorine bleach to one quart of water. Pour one to two cups of treatment chemical over the sewage each time you use the latrine. **WARNING**: *Don't use powdered bleach — it is poisonous.*

√ Sewage can be burned outside in a metal container. After burning, bury it in the ground.

√ Sewage can also be temporarily held in a plastic container or a trash bag until it can be burned or flushed down a working toilet. After each use, pour one or two cups of treatment chemical over the waste.

How to Make a Latrine

√ To make a latrine: dig a hole 200 feet away from any wells, streams, lakes, or water sources. Dig a hole 3 to 6 feet deep and 2 to 4 feet wide. After each use, pour one to two cups of treatment chemicals, used to kill germs and control odors, over sewage. Ash from a fire can be placed over sewage to control odors. Then place about one inch of dirt on the top of the waste to keep away any flies and disease spreading rodents. When the waste reaches six inches from the ground level, fill the hole with dirt and dig a new latrine.

24

Shelter

How to Build a Quick Shelter Inside a House for Warmth

If it is cold and the power fails or the furnace or heater is not working, you can make a shelter inside one room in your house.

- Use desks and small tables for the shelter frame. All four sides, the roof and the floor should be covered with the insulating materials mentioned below.

- Use mattresses, blankets, towels, pillows, sheets, coats, or clothing to make walls 15 inches or thicker.

- Make the shelter compact with only a small space for each person. This way your body heat will help warm the shelter. See pages 20 & 21 for tips on keeping warm.

Quick Backyard Shelter for Warmth

- A shelter can be built very quickly by shaping a piece of canvas, plastic, a tarp, or blankets into the shape of a tepee, square box or other enclosed shape. Place a portable wood burning stove inside or keep a small fire burning. This will keep it warm inside. *NOTE: You can make your own stove with a metal box or a large metal can or a similar container. Cut a hole in the side of the box to insert fuel and a vent hole on top for the smoke to get out. Keep an open space for fresh air to come in and make a vent for the smoke to escape from the stove through the top of your shelter. If you are using an open fire, make a hole in the top of the shelter. The hole should be eight feet or higher off of the ground. WARNING: Do not use galvanized metal for your wood-burning stove because it can emit poisonous fumes.*

- Sleep on cots, mattresses, or place 15 inches or more of insulating materials underneath you. Read below for details on Insulation.

General Shelter Building Information

- Look at the shelter illustrations on pages 48 to 51.

✓ Pick out the shelter that you think is best suited to your situation or improvise as needed.

✓ Begin building your shelter five hours or more before sunset.

✓ Choose a safe location. Do not build near avalanche or flood prone areas such as dry stream beds. A good location would be in a dry, sunny and flat area near a source of water and firewood.

✓ A fire should be built next to the shelter or inside the shelter. Fires give warmth and are also very comforting. Read the chapters on HEAT and

TEMPERATURE: How to Get or Stay Warm (page 20) and How to Build a FIRE (page 9) for more information. Also, see the illustrations on pages 43 & 44 and 48 to 51.

√ When it is cold and there are no available materials to build a shelter, you can heat rocks in a fire for several hours or more. *CAUTION: Don't use round river rocks — they can explode.* Bury the rocks about six inches deep, directly underneath where you will be sleeping. Place nonflammable cushioning material over the ground and sleep on top of this. You should be able to keep warm with this set up.

• A type of shelter that can be built very quickly is the Squirrel's Nest shelter. See the illustration on page 48. Use dry grass, leaves, car seat foam, or old pine needles for the filling.

√ Tepees are excellent for cold weather. Use a blanket, tarp, or a piece of plastic for the covering. Keep a small fire going inside for warmth. You will need six to eight feet of room overhead with a hole at the top to vent the smoke. See page 51 for an illustration.

√ If you build a cold weather shelter using a tree, pick a tree without snow on the branches or knock the snow off the branches before using it for a shelter. See the illustration on page 51. Snow in branches will fall on you when the shelter warms up.

√ Do not lay or sleep on the cold ground. Place lots of insulating materials on the ground first. Read page 65 about Insulation for details.

How to Tie Materials Together

√ Branches can be tied together with thin strips of bark, wire, or string. You can also cut notches to hold the branches together. If a branch is too long to use, it can be burned or cut in half to make it shorter.

Shelter Covering

√ Mud mixed with grass or leaves can be spread over the surface of a shelter to waterproof it. Birch tree bark makes an excellent shelter covering. Peel off large strips of bark by cutting downward in a spiral pattern on the trunk of the tree.

√ Grass can be tied in bundles and used for part of your shelter covering. Place the bundles up against the outside of the shelter with the grass blades pointing up and down (not sideways), then tie them in place. This should be done with several layers of bundles on top of each other.

Fire Information for Shelters

• Read the chapter on How to Build a FIRE (page 9) for more complete information.

• Your fire should be small so you can get close to it and stay warm.

• A reflector should be built in back of the fire with rocks or stacks of wood to reflect the heat from the fire back towards you.

- If you are going to build a fire inside your shelter, then you will need plenty of room overhead — at least six to eight feet. You will need a hole at the top so the smoke can escape and a vent hole for air.

- You can also heat rocks as explained previously in General Shelter Building Tips. Place two or three of the heated rocks just inside the door of your shelter. This will keep the inside of the shelter warm for a couple of hours. Replace the rocks when they cool down with fresh, hot rocks from the fire. Build a small shelter with a tiny opening for this method. *CAUTION: If you use this method, make sure that there is nothing near the rocks that could catch on fire. Place something around the rocks so that you don't roll on them or burn yourself while you are sleeping.*

Insulation Information for Shelters

- Insulation is a very important component of a shelter. The ground will make you cold very quickly.

- To help prevent your body from getting cold on the ground, you should have 15 inches or more of insulation or six inches of foam between you and the ground. Foam from car seats may be used. The insulation should be light, dry materials. Good insulation materials are pine boughs, loose dry grass, leaves, cattail fluff, pine branches, tied up bundles of dry grass, old dry pine needles or slightly crumpled sheets of newspaper. *CAUTION: If you are using the buried hot rock heating method, you should use insulation that cannot catch on fire or protect the insulation from catching on fire by placing rocks or non-burning materials around it.*

25

Shoes

Footwear is very important — just ask any experienced long distance hiker. Wear two pairs of socks to help prevent foot blisters on long hikes. There are several types of emergency footwear that can be self made. Below are some examples:

Emergency Snowshoes

Snow shoes can be made from green, tree branches tied with string or thin strips of green bark. See the illustration on page 52 for details.

Emergency Winter Shoes

Cold weather emergency shoes can be made with household materials such as towels, plastic and newspapers; a canvas-type material is good for dry weather conditions. Cut a square 30 inches by 30 inches out of the material you selected. *NOTE: For cold, wet weather, plastic bags (grocery bags are best) or sheets of plastic can be used to keep the feet dry.*

- Place a 30 inch by 30 inch square of material on the ground.
- Then place ten layers of newspaper that have been separated first, slightly crumpled, and then put back together on top of the selected material.
- Place a towel, a piece of foam (available from mattresses or car seats), or other dry insulating materials like dry grass, or dry leaves on top of the newspaper and/or foam. *NOTE: towels and foam are the best choices.*
- With dry socks and shoes on, place your foot on top of all layers of these materials and tie the entire bundle around your upper ankle.
- If you are using a plastic covering for wet weather, stop occasionally and remove the plastic to let the built up moisture escape.

Emergency Warm Weather Shoes

- To make warm weather shoes (emergency sandals), find a piece of regular car tire or a thick, hard piece of plastic or similar materials.
- Cut out a piece of this material that is about one inch bigger than your foot all the way around. Do this by tracing the outline of your foot on the material.
- Punch one hole at the front and two holes at the back. Next, lace a long piece of string (36 inches) or a strip of cloth through the holes just like on a sandal. This is to hold your foot in place.

26

Signals

Signals that are known by rescuers all over the world are shown on page 54. Pick out the signal or signals that are relevant to your situation.

NOTE: *Three of anything (known as SOS) — whistle blows, fires, etc. — is the international signal for "Need help!"* (see page 54 for morse code SOS.)

✔ Three fires in a triangle is an emergency signal known all over the world. If you are stranded, make three separate piles of wood. Keep them ready to be set on fire. If you see or hear a plane approaching, light them on fire.

• A spare tire from the trunk can be lit on fire with a flame (match, lighter, etc.). This can be used to keep you warm and also be seen as a signal for many miles. You can also take a tire off the car and burn it if necessary. Just remember, if you remove a tire from the car and burn it after you've burned your spare tire you will no longer be able to drive the car. **CAUTION:** Tire smoke is harmful, avoid breathing it. Also, once the tire is burning, it is almost impossible to put it out.

✔ If you need to be rescued, a single, large, smoky fire should be kept burning at all times in a survival or disaster situation. You can make a fire smoke by placing green leaves or green grass on the fire.

✔ When you make a signal, it must contrast with its surroundings. This is the most important element of a signal. A good example of contrast would be dark branches placed on light colored sand or snow. Make the size of your signal letters large: at least 25 feet long (see top of page 54). Signals should be placed in a clearing so that they can be easily spotted from the air.

✔ Destroy your signals when you are rescued.

✔ If you are stranded by the roadside or your car is broken down, you can signal to a passing trucker to call for help. Use one hand to point at the truck driver and get the truck driver's attention. Use your other hand to make the form of a telephone receiver and mouth the words, "Help." Most truckers can radio to the nearby town on their CB radio (Channel 9) to have help sent to your location. See an example in the Illustrations on page 55.

• Raising both arms straight up is a sign for "Help." Raise one arm to say "Hello, everything is okay." (see page 54)

27

Supplies and Advanced Preparation

This is a checklist for emergency and disaster supplies. Getting prepared before a disaster could mean the difference between life and death. Don't forget to keep emergency supplies (at least a 72-Hour Emergency Kit and water) in each car and at your work place or school. Some of the large quantities recommended here are in the event of a long term major disaster. Most of these items can be purchased from the company listed at the bottom of page 100.

Standard Disaster Supplies - For ALL Disasters

These are the minimum supplies that should be stored.

72-HOUR EMERGENCY KIT Containing the following:

√ A comprehensive emergency disaster instruction book

√ Food: 10 days to 3 weeks supply of food needed. ~~Dehydrated cooking meals are best. Coast Guard approved food bars are also good. *NOTE: Don't use MREs (Meals Ready to Eat) for long-term storage in your supplies. They are easily damaged by freezing and other extremes in temperature.~~*

√ Water: Lots of it. Retort pouches are best because they aren't damaged by freezing and extremes in temperature (see the chapter on *WATER* (page 79) for additional information about water storage)

√ Water purification filter and/or water purification pills (iodine based)

√ Water carrier and wash basin

√ Tent

√ Sanitation supplies including: 25 kitchen size trash bags and ties to use for a toilet and sewage treatment chemicals (the powdered type is best)

√ Toiletries: Toothbrush, toothpaste, soap, razor, shampoo, female needs, etc.

√ First aid kit and a comprehensive first aid book

√ Portable Radio: Hand crank type that needs no batteries

√ Heat Source and Extra Fuel

√ Candles and Light Sticks

√ Fire starter: Matches (waterproof) in a waterproof container

√ Blankets: Wool and/or a non-temporary metalized space blanket are best

√ Clothing: Thermal underwear, hats, sturdy shoes, waterproof mittens, etc.

√ Cash in one dollar bills or pre-1965 silver coins & quarters for phone calls

- Fishing Kit, hooks, lures, 200 yards of heavy line, etc.
✔ Heavy duty leather work gloves
✔ ~~Swiss Army~~ knife (~~genuine if possible~~)
✔ Compass and maps
✔ A sewing kit and three or four packs of dental floss for tying things *and paracor*
- Photocopy of important documents: birth certificate, drivers license, etc.

Standard Disaster Supplies

☑ CHILD NEEDS: foods, formula, bottles, diapers, clothing, games, medicines

☑ DISASTER/EMERGENCY BOOK with first aid instructions

☑ WATER: 30 gallons or more per person (see the chapter on WATER for more information on water storage)

☑ WATER STORAGE IN CAR: ~~retort pouches (foil pouches) will not be damaged when it freezes while being stored in your car trunk, etc.~~

☑ PORTABLE WATER FILTER and WATER PURIFICATION PILLS (iodine)

☑ WATER PURIFICATION PILLS: 1 bottle for each person

☑ FOOD STORAGE: for each person, store at least a one year supply of food. (listed below is a recommended 1 year's supply of food storage.) Dehydrated long term food storage should be packed with nitrogen or CO_2 (to remove oxygen). Buy the type of food storage that is sealed in #10 metal cans (approximately gallon size) or in 5 gallon plastic buckets.

- 75 pounds dry milk

- 325 pounds of a variety of grains: wheat, corn, oats, barley, rice

- 60 pounds of a variety of legumes: peas, pinto beans, kidney beans, lentils

- 50 pounds of sprouting seeds (alfalfa is recommended.)

- 60 pounds of honey (optional)

- 40 pounds of fat

- 25 pounds of salt

- Vitamins and minerals - 1 to 2 year's supply

- Wheat Grinder

☐ FOOD STORAGE COOKBOOK: The New Cooking with Home Storage By Vicki Tate or Marlene's Magic with Food Storage By Marlene Petersen. These books are available from the supplier at the bottom of page 100.

☐ GARDEN SEEDS: store the non-hybrid type — buy enough seeds to grow at least a year's supply of food for each person. Check packaging date for current year.

☑ EMERGENCY FOOD BARS

☑ EMERGENCY HEATER/STOVE: wood burning or propane.

☑ ELECTRIC GENERATOR and FUEL

☑ FIRE STARTER: butane lighter, matches (waterproof), magnesium stick

☑ ABC TYPE PORTABLE FIRE EXTINGUISHER

☑ CLOTHING: insulated mechanics overalls or ski overalls are good to have. Wool hats, pants, and shirts will keep you warm. Winter coats, scarves, gloves and mittens should also be stored. It is best to wear several layers of clothing to keep warm. You can add or remove individual layers according to your needs.

☑ SHOES: sturdy shoes that are suitable for hiking and walking should be stored. Waterproof shoes such as hiking boots, quality athletic shoes, and good winter boots are priceless when needed.

☑ BLANKETS and FOAM SLEEPING PADS: wool or wool blend blankets are best because they can keep you warm even when they are wet. Store enough blankets to keep everyone warm if there is no heat available.

☑ CANVAS WALL TENT or TEPEE: small, compact type with a waterproof bottom

☑ ROLLS of PLASTIC SHEETING, TARPS for temporary windows & shelters

☑ EMERGENCY LIGHTING: light sticks that you bend to activate are safe to use when other light sources may be dangerous to use. Flashlights and batteries, propane lantern and fuel, oil lamp, candles, or matches are also good to store.

☑ RADIO: the hand crank type radio that never needs batteries is the best kind. These are available from the supplier listed on the bottom of page 100.

☑ SANITATION: portable toilet and treatment chemicals, toilet paper and plastic bags

☑ SOAP: antibacterial soap (10 bars per person)

☑ FEMALE NEEDS: sanitary napkins also make good emergency first aid bandages and compresses

☑ CHILD NEEDS: foods, formula, bottles, diapers, clothing, games, medicines

☑ COMPREHENSIVE FIRST AID BOOK

☐ FIRST AID KIT: A first aid kit should include at least the following items:

✔ thermometer	✔ latex gloves	✔ soap
✔ band aids	✔ gauze/bandages	✔ elastic bandages
✔ medical tape	✔ rubbing alcohol	✔ cotton balls/swabs
✔ antibiotic ointments	• tourniquet	✔ scissors
✔ razor blades	• splints	✔ tweezers

✔ diarrhea medicine   • ipecac syrup

✔ aspirin and other pain/inflammation relievers

- oral antibiotics such as penicillin, tetracycline

WARNING: Always check your expiration dates on medicines and supplies. Tetracycline is poisonous after it has expired. Also, don't give tetracycline to children under the age of eight because it stains their teeth.

☐ For juvenile, Baby-sitter information check list and first aid instructions (laminated) send $1 to:

Council on Family Health
225 Park Avenue South, New York City, NY 10003

☑ MEDICINES: Store any regularly used prescription medications in addition to commonly used over-the-counter (OTC) medicines.

☑ DENTAL EMERGENCY KIT

☑ PROTECTION DEVICE: Gun, pepper mace, electric stun gun, crossbow. Knife ~~Read the book The Truth About Self-Protection, Bantam Books®.~~

☑ GAS WRENCH or CRESCENT WRENCH: To shut off natural gas valve

☑ GLOVES: Heavy duty work gloves

☑ KNIFE: Swiss Army type or a high carbon steel hunting knife

☑ ~~DIAMOND TYPE~~ KNIFE SHARPENER

☑ SHOVEL and PICK

☑ TIRE IRON: (for moving things after a disaster), PLIERS, WIRE CUTTERS and BOW SAW

☑ ROPE (200 feet nylon) and ~~SILVER~~ DUCT TAPE (2 large rolls), paracord

☑ EYE GLASSES: Extra pairs

☐ ANTIFREEZE: The nontoxic type used in RVs is preferable, but regular car antifreeze can be used for protecting toilets or plumbing from freezing during a power failure. *WARNING: Regular car antifreeze is poisonous to humans and animals.*

☑ SUPER HEAVY DUTY LARGE TRASH BAGS (50 to 100)

Complete List of Disaster Supplies

To be really well prepared, you should have all of the things on the following list in addition to the supplies listed above. Extra supplies can be traded or used to help others in a disaster. Don't forget to also keep supplies in each car.

☐ BOOKS: "Where There Is No Doctor", "Where There Is No Dentist", "Nuclear War Survival Skills" and "Field Guide to Edible Wild Plants"

☑ WEATHER INFORMATION RADIO: this is a special radio that gives constant weather reports. The better models come with an alarm that should automatically go off in the event of severe weather.

☑ WATER PUMP

☑ SAND BAGS

☑ POTASSIUM IODIDE: 1 bottle for each person. (This substance helps to protect the thyroid gland in case of nuclear accident or nuclear war.)

☑ RADIATION FALL OUT METER and radiation dosimeter *or badge*

☑ PORTABLE WALKIE TALKIES & extra batteries

☑ SOLAR CHARGING LIGHT: available at some hardware stores

☑ HAND WARMERS: air activated

☑ ~~SEVERE~~ COLD WEATHER CLOTHING: ~~the Expedition Series Suit from Northern Outfitters sells for about $1,100; phone 1-800-944-92~~76

☐ CLOTH for manufacturing your own clothing

☑ SEWING KIT and extra thread

☐ SNOW SHOES

☑ INSECT REPELLENT

☑ POST HOLE DIGGER

☐ SPOOLS OF WIRE

☑ BINOCULARS

☑ WHISTLE

☑ PRESSURE COOKER and/or DUTCH OVEN

☑ ~~DRUM OF~~ FUEL ~~55 GALLONS oil, kerosene~~

☐ NEWSPAPER LOGS FOR FUEL. **Directions:** *Divide newspapers into half page (folded) sections. Add a tablespoon of detergent to a bucket of water. Soak folded sections for 5 minutes. While still wet, roll the sections on a one inch diameter rod. Squeeze out the excess water. Slide roll off rod and stand on end to dry. Rolls are ready to use as fuel when completely dry.*

Where to Buy Disaster Supplies

There are companies which specialize in disaster supplies. Write and ask for a free catalog.

> See the bottom of page 100 for the name and
> address of a company that sells disaster supplies.

28

Tidal Wave

TSUNAMI — or TIDAL WAVES as they are more commonly called, kill many people around the world. They can travel up to 500 miles per hour.

Advance Preparation

Buy the Standard Supplies that are vital for any disaster. These are listed in the SUPPLIES - ADVANCE PREPARATION (page 68) chapter under the heading Standard Disaster Supplies. It is very important to read the entire chapter. Most of these items can be purchased from the company listed at the bottom of page 100. In addition to the Regular Disaster Supplies, you should stock the following Special Supplies to be prepared for a Tidal Wave:

☐ SANDBAGS   ☐ BOAT

☐ LIFE JACKETS   ☐ WATER PUMP

Preparation Tasks and Skills

☐ Learn in advance where your community evacuation routes are.

☐ Learn in advance the location of your local disaster shelters. Practice driving to one at least once or twice a year.

☐ Learn what the bomb/disaster warning sirens sound like.

☐ Read through this entire book so that you are more familiar with other disasters and the necessary advance preparations and supplies needed for each type of disaster.

Tidal Wave Warning

THIS MEANS A TIDAL WAVE IS COMING!

If you hear a warning, go immediately to higher ground! Do not return home until 24 to 36 hours after the tidal wave hits unless officials tell you that it is safe to return home sooner. The reason for this is that smaller tidal waves frequently follow the first wave and have killed many people who thought they were safe after the first wave struck.

Take the following precautions

(only if the officials tell you that there is enough time)

☑ Turn off the gas, electric and water supply to the house. *Read the chapter titled TURNING OFF THE GAS, ELECTRICITY and WATER (page 78).*

☑ Find your 72-Hour Emergency Kit to take with you. *See the SUPPLIES - ADVANCE PREPARATION chapter* (page 68) *for details.*

☑ Bring water, food, blankets, a first aid kit and shelter with you.

After the Tidal Wave

√ Help the sick and elderly, or anyone who may be injured or has other special needs.

√ If there is water in the house or if the house has been damaged, have the gas, electricity and water turned off before you enter the property. Read the chapter TURNING OFF THE GAS, ELECTRICITY and WATER (page 70) for details.

√ Do not drink the water from the faucet until the radio or TV broadcaster says it's okay to drink from safely. Use bottled or storage water instead. See page 79 for purification information.

√ Do not use any open flames, matches, candles, or lanterns until you are sure the gas is not leaking. Use a chemical light stick instead.

√ Don't enter unsafe damaged homes or buildings.

√ Contact your insurance agent if there are injuries or property damage.

29

Tornadoes

Tornadoes usually develop out of severe thunderstorms. They have very high winds and extremely low air pressure. Tornados are very powerful funnel shaped storms, they are extremely loud. Many people say they sound like a train — they can destroy a house or a building in a matter of seconds.

Advance Preparation

Buy the Standard Supplies that are vital for any disaster listed in the SUPPLIES - ADVANCE PREPARATION (page 68) chapter under the heading Standard Disaster Supplies. It is very important to read the entire chapter. Most of these items can be purchased from the company listed at the bottom of page 100.

In addition to the Regular Disaster Supplies, you should stock the following Special Supplies in case of a Tornado:

☑ WEATHER INFORMATION RADIO WITH A SEVERE WEATHER ALARM: these radios give information on the weather constantly and should sound an alarm if a tornado is in your area

☑ CROW BAR: a large one for moving junk, prying boards

Preparation Tasks and Skills

☐ Learn in advance where your community evacuation routes are.

☐ Learn in advance the location of your local disaster shelters. Practice driving to them one to two times a year.

☐ Learn what the bomb/disaster warning sirens sound like.

☐ Read through this entire book so you are more familiar with other types of disasters.

Tornado Watch

A tornado watch tells you that a tornado is likely to develop in the area. The radio or TV will give official information and instructions during tornado weather.

☑ To be extra safe, you may want to go to a shelter when they announce a tornado watch before an actual tornado develops.

☑ Listen to your radio or TV for further information and instructions.

☑ Keep a watch for approaching tornadoes. They are shaped like a funnel and are very loud. If you see one coming, take shelter immediately!

Tornado Warning

THIS IS VERY SERIOUS. A tornado has been sighted in the area. Take shelter immediately!

- ☑ If you are in a car, go to the nearest shelter. If a shelter is not available, read the paragraph below on being OUTDOORS.

- ☑ If you have a basement, go to a corner that is away from any windows. If possible, get under a desk, table, or a similar object to help protect yourself. Do not get under a section of floor that has a refrigerator or other heavy object above. These could fall through the floor and land on top of you if the tornado strikes your home.

- ☑ If you are in a home or building without a basement, go to the lowest floor and get in a closet or a bathroom. If these are not available, get under a sturdy doorway or in a center hallway.

- ☑ If you are in a mobile home, leave and go to the nearest shelter. Mobile homes are very dangerous to be in during a tornado.

Outdoors

- ☑ Get away from power lines, buildings, trees, and other objects that could hurt you when the tornado hits. Keep low to the ground and cover your head with your hands.

- ☑ If you are driving, drive in the opposite direction or at right angles to the path of the tornado.

- ☑ If you are in a car and and the tornado is coming right at you, pull under an overpass and climb up the embankment and position yourself in the nook under the overpass

After the Tornado Passes

- ☑ Help the sick, injured and the elderly. Also assist others with special needs.

- ☑ Listen to the radio or TV for official instructions and information.

- ☑ Do not enter unsafe damaged homes or buildings.

- ☑ Never use an open flame inside a dwelling (no matches, candles, or lanterns) until you are positive that there are no gas leaks or spilled flammable chemicals. Instead, use a chemical light stick that you bend to activate.

- ☐ Don't use the telephone unless you have a serious emergency; making unnecessary calls will jam lines.

- ☑ Beware of broken gas lines and downed electrical lines.

- ☑ Don't plug in any wet appliances such as refrigerators if there is water in the house. Wait until everything is dry. You may want to have a home inspector check your home first to be sure it is safe.

Don't drink water from the tap until the TV or radio broadcasters say it is safe to drink. It could be contaminated with sewage or chemicals. Drink your stored or bottled water instead. See page 79 for more information.

 Call your insurance agent if there are injuries or property damage.

30

Turning Off Gas, Electricity and Water

Gas

To turn the gas off in your house, go outside and find the gas meter. It is usually on the side or back of the house. Locate the shut off valve on the gas pipe. Then with a wrench turn the valve 1/4 turn in either direction (this is a 90 degree turn). The valve should now be going across the pipe (perpendicular). The gas is now off. See the illustration on page 58 for a visual explanation. *Caution: Your gas company may not allow you to turn your own gas valve back on once it has been shut off.*

Electricity

To turn off the electric service, find the circuit breaker or fuse box. It is a square metal box. See illustration on page 57.

CIRCUIT BREAKER: Push the large main breaker switch to the off position. See the illustration on page 57.

FUSE BOX: Locate the large main fuse. Grab the handle and remove it from the fuse box. See the illustration on page 57.

Water

There are two ways to turn off your water:

INSIDE: Look in your house and find out where the water pipe enters — this is usually on a lower level. Find the shut off faucet valve and turn it clockwise all the way to the right. The water should now be off. See the illustration on page 56.

STREET WATER VALVE OUTSIDE: *(Caution: Make sure you're not shutting off other people's water supply. ie; apartments, duplexes, etc.)* Find the water meter — it is usually located by the curb. To turn off the water, first open the lid take a wrench and turn the valve 1/4 turn (this is a 90 degree angle). Some of these only turn in one direction. The valve should be going across the pipe (perpendicular). The water is now off. See the illustration on page 56.

31

Water

Having water is more important than having food in a short term emergency.

WARNING: *Never drink water from your radiator; antifreeze is very poisonous.*

✓ After an earthquake, flood or any major disaster — do not drink the tap water until authorities on the TV or radio tell you that it is safe to drink. It may be contaminated. Drink bottled or storage water instead.

✓ All water should be purified before drinking in a disaster or survival situation, even water that comes from a mountain spring. This is because the water could be contaminated with Giardia, sewage, or other contaminants. Steps to purify and store water are covered below.

✓ Drink lots of water in a survival situation. A good amount for an adult to drink is at least ½ gallon a day (1 gallon a day in hot weather). You are drinking enough water when you urinate 2 pints of liquid a day (1 pint a day is a bare minimum to urinate). Also the color of your urine should be light yellow or clear. **NOTE:** *B vitamins and some medications will darken the color of urine.*

✓ Ration your water sensibly. People have died of dehydration with water still in their canteens.

✓ Remember to "ration your sweat and not your water." Do this by using less energy. If there is little or no water available, keep activities to a minimum and stay out of the sun. If you have to walk, do it at night to minimize water loss through perspiration.

✓ If you have food and no water, you should eat very small amounts of food or wait until you have water before you eat food. Your body uses water to digest food. Having no water and then eating food would make you even more dehydrated.

Signs of Dehydration (Low Body Water)

- Dark colored urine
- Thirst
- Dry mouth and inside of nose

- Tired
- Fatigue
- Sleepiness

- Fever
- Headache
- Dizziness

- Loss of appetite
- Abdominal skin that will not stretch

Storing Water for Emergencies

✓ Store 30 gallons or more (as much as you can) for each person.

✓ You can purchase special water storage containers from camping and preparedness suppliers such as the one listed at the bottom of page 100. Clean, 2-liter plastic soda bottles make excellent water storage containers. CAUTION: Don't store water in plastic milk jugs because these containers don't store well and the bottles rupture. Also, don't store water in bleach jugs because the plastic contains harmful chemicals.

√ Treat water for long term storage by adding plain liquid chlorine bleach with sodium hypochlorite as the only active ingredient. *WARNING: Never use powdered bleach because it is poisonous. Also, NEVER use any liquid bleach that has added softeners or scents, or has color safe additives because these are harmful and poisonous.* Bottled water needs no treatment

Liquid chlorine bleach should be added to water intended for storage in the following amounts:

WATER	BLEACH
2 liters	9 drops of bleach
1 gallon	16 drops of bleach
5 gallons	1 teaspoon of bleach
25 gallons	5 teaspoons of bleach
55 gallons	3 to 4 Tablespoons of bleach

• Fill the water level to the top of the container without leaving any air space at the top. This helps slow bacteria growth.

Emergency Water Filter

Caution: You still need to purify water after filtering (see instructions in this chapter).

You can make an emergency water filter by placing alternating layers of sand, charcoal, grass, dirt and tissue paper (if you have any tissue paper) in a container that is about 12 inches or so deep. Punch a hole in the bottom for the water to flow out. *NOTE: The lowest layers should be sand and charcoal.*

Purifying Water for Drinking

All water should be purified and filtered (if possible), especially if it came from the outdoors. Read the directions below for details.

√ Boiling for 20 minutes is the best way to purify water.

√ Commercial water purification pills can be used. Follow the directions on the container.

√ Plain liquid chlorine bleach can be used to purify water. Add three drops of plain liquid bleach per quart for clear water and four drops of plain liquid bleach per quart of cloudy or muddy water. Let sit 30 minutes at normal temperature (up to 1 hour if very cold) before drinking. *WARNING: Never use powdered bleach or any bleach that has any ingredient other than* sodium hypochlorite in it — other ingredients are poisonous. Also, do not use scented or color safe liquid bleaches.

✓ After a nuclear war or a nuclear accident, the water will probably be contaminated with radiation. You can remove 99% of the radiation by pouring the water though a large bucket that is filled with dirt. **DIRECTIONS:** Punch three small holes in the bottom of the bucket for the water to drain out. Next, place about two inches of sand and small pebbles in the bottom of the bucket. Then fill the bucket with regular dirt. Pour the water through the bucket.

Finding Water Indoors

If you are indoors, there are several ways to get water when the regular supply for your community has been damaged or contaminated.

✓ You can drink the water in your toilet tank — the upper part of your toilet and NOT THE BOWL. **WARNING:** Do not drink the water from your toilet tank if your toilet has a disinfectant, deodorant, or a chemical that colors or scents the water.

✓ You can drink the water in your water heater. First turn off the gas and water supply to the tank. Next, turn on the small valve or faucet at the bottom of the water heater. **NOTE:** You may need to open the valve on the top of the water heater before the water will flow out.

Finding Water Outdoors

When you are dehydrated, or, in other words very thirsty, do not eat snow to obtain water — it must be melted first. You use up energy and body water to melt the snow. If you eat snow, you will become even more dehydrated than you were before you ate the snow. Melt snow by placing a container of snow next to a fire or in the warm sun.

Below are several methods for finding, gathering and purifying water. Muddy or unclear water should be filtered and then purified. Read the details on how to make an Emergency Water Filter at the end of this chapter.

✓ At the ocean or near a muddy stream, go about 50 feet away from the water's edge or use the high tide mark for the ocean. Dig down until water seeps in. This is a simple filtering process. You still need to purify the water (if possible).

✓ If it is cold and there is snow available, you can make a snow melting device: Place snow in a plastic tarp, bag, bucket, or a similar container. Keep this container near a heat source such as a fire or a stove. Wait for the snow to melt. Next, let the water get warm after it melts because cold water chills your body.

✓ Collect rain water by setting out plastic sheets or buckets before or during a rain storm. Also collect all the rain water that you can after a storm. Muddy water from puddles can be filtered and purified as explained in this chapter.

- You can make a solar bag for water collection. **NOTE:** *This only gathers a small amount of water.* **Directions:** *Get a plastic bag or a plastic sheet. Take the bag or sheet and fill it with plants that are not poisonous such as grass or dandelions. Tear up the plants before you place them in the bag. Tie the bag shut and place it in the sun. Water will condense on the bottom.* **CAUTION:** *Many regular trash bags are coated with small amounts of insecticides (bug killer) which can be harmful to humans.*

√ Check stumps, depressions in rocks and cliffs for pockets of water. Also check low lying areas for streams or ponds. Remember that water seeks the lowest level that it can find. Pools of water from rain can be found on large rocks, on top or on the sides of rock cliffs and on rocky type mountains. Water can sometimes be found at the base of cliffs if you dig down a couple of feet.

√ Soak yourself in a stream or lake: your body will absorb water.

√ Mop up dew (the morning time is best) from grass, rocks, tree, or leaves with a rag, cloth, or shirt. Then squeeze the water out and purify before drinking (if possible).

√ Dig in the outside turns of dry creek beds to reach water. **WARNING:** *Save this for a last resort because flash floods have been known to come without warning and sweep people away, especially in desert areas.*

32
Winter Storms

Winter storms can come on suddenly and cause lots of problems. Listen to a portable radio or the TV for information and instructions about the storm.

WARNING: *Many people die from heart attacks as they shovel the snow from a storm — so don't over exert yourself.*

The chapters titled HEAT and TEMPERATURE: How to Get or Stay Warm (page 20) (which includes sections on Frostbite and Hypothermia), POWER FAILURE - BLACKOUT (page 39) and SHELTER (page 63) can be very useful when the power fails or when it is cold and storming. Also see the chapters on CLOTHING (page 4) and SHOES (page 66) for details on protecting yourself from the weather.

Advance Preparation

Buy the Standard Supplies that are vital for any disaster. These are listed in the SUPPLIES - ADVANCE PREPARATION chapter (page 68) under the heading Standard Disaster Supplies. It is very important to read the entire chapter.

In addition to those supplies, you should stock the following Special Supplies in case of a winter storm. Most of these items can be bought from the company listed at the bottom of page 100.

- ☑ WEATHER INFORMATION RADIO with a severe weather alarm
- ☑ SNOW BLOWER
- ☑ WINTER CLOTHING: boots, ski suit, gloves, hat, scarf, insulated mechanic's overalls
- ☑ BLANKETS, QUILTS and FOAM SLEEPING PADS
- ☑ BAGS OF SAND or SNOW MELTING CHEMICALS
- ☑ WOOD BURNING STOVE or other emergency heater with a good supply of fuel such as firewood

Preparation Tasks and Skills

- ☐ Learn in advance where your community evacuation routes are.
- ☐ Learn in advance the location of your local disaster shelters.
- ☐ Learn what the bomb/disaster warning sirens sound like.
- ☐ Buy a CB (permanent or portable) for your car.
- ☐ Read through this entire book so that you are more familiar with disasters.

Winter Storm Warnings and What They Mean

√ **TRAVELERS ADVISORY:** Snow or ice is expected. Travel and visibility may be difficult.

√ **WINTER STORM WATCH:** Severe winter weather is possible. Stay tuned to TV or radio for more information.

√ **WINTER STORM WARNING:** Heavy snow or freezing rain is expected.

√ **BLIZZARD WARNING:** Winds of 35 miles per hour or more. Snow and temperatures 20°F or lower are expected.

√ **SEVERE BLIZZARD WARNING:** Winds 45 miles per hour or more. Snow and temperatures 10°F or lower are expected.

Winter Storm Watch

A winter storm watch tells you that there is a very good possibility that severe storms and/or winter weather is expected in your area.

☑ MONITOR THE TV OR RADIO for weather information.

☑ ORGANIZE YOUR EMERGENCY SUPPLIES: read the Advance Preparation section above and also in the SUPPLIES - ADVANCE PREPARATION chapter (page 68).

☐ Read the list of needed supplies at the beginning of this chapter and also read the chapter SUPPLIES - ADVANCE PREPARATION (page 68).

☑ STAY AT HOME or CLOSE TO HOME if possible.

☑ Find your shovel and snow melting chemicals.

☑ Make sure everyone has good winter boots, blankets and winter clothes.

Winter Storm Warning

A winter storm warning is a warning that dangerous snow storms and/or sleet and freezing rain are expected in your area. Get prepared by doing the following:

☑ MONITOR THE TV OR RADIO for weather information and official instructions and information.

☑ ORGANIZE YOUR EMERGENCY SUPPLIES. Read Advance Preparation list at the beginning of this chapter and also read the SUPPLIES-ADVANCE PREPARATION chapter for a list of needed supplies.

☑ GET A GOOD ALTERNATE HEATING SOURCE and fuel if you do not already have them.

☑ STAY AT HOME or CLOSE TO HOME if possible.

☑ GATHER A SHOVEL AND SAND, also snow melting chemicals.

☑ FIND BLANKETS, WINTER CLOTHES and WINTER BOOTS (one pair for each person)

If You Are Stranded in a Car

☐ Read the other chapters in this book that might help such as: PRIORITIES (page 59), HEAT and TEMPERATURE (page 20), FROSTBITE (page 22), HYPOTHERMIA (page 22), LOST or STRANDED (page 31), HOW TO BUILD A FIRE (page 9), SHOES (page 66), SIGNALS (page 67), SHELTER (page 63).

☑ If you have a CB Radio installed in your vehicle or carry a portable CB, you can call for help on Channel 9 — the police regularly monitor this channel. *or use cell phone*

☑ Pull off to the side of the road.

☑ If there is traffic, you can signal to a passing truck driver to call for help. Use one hand to point at the truck driver and get the driver's attention. Use your other hand to make the form of a telephone and mouth the words, "Help." Most truckers can radio to a nearby town on their CB radio (channel 9) for help. See the illustration on page 55.

☑ It is usually best to stay with your car, unless you can see a building or a shelter within safe walking distance.

☑ Tie a bright colored cloth on your antenna so rescuers can find you. Also place a colored cloth on the snow next to your car, (with something heavy on top of it to hold it down) so that rescuers searching from the air can spot you.

☑ If there is more than one person in the car, have one person sleep and the other keep watch for rescue crews. Take turns.

☑ When it gets very cold, open a window an inch or two and turn on the engine. Run the heater for a few minutes until it is warmer. *WARNING: First make sure that the muffler is not clogged with snow. A clogged muffler may cause carbon monoxide poisoning.*

☑ Car seats have foam inside of them which can be ripped out and tied around your body for warmth. First make a hat from the foam because you lose 50% of your body heat from your head. Also cover your feet and hands because they also lose a lot of heat. See the chapter on CLOTHING (page 4) for information on improvising clothing in cold weather.

NOTE: Most car insurance companies can add a towing coverage clause to your policy for minimal cost. You might also consider joining an auto club such as AAA®.

33

Where to Call in an Emergency

Fill In with Local Phone Numbers

Emergency _____

Police / Sheriff _____

Highway Patrol _____

Weather and Road Information _____

Doctor _____

Ambulance / Rescue _____

Hospital _____

Poison Control Center _____

Relative _____

Neighbor _____

Other _____

Other _____

34

Emergency Numbers and Addresses

Emergency Shelters (phone and address)

_____

_____

_____

_____

_____

Out of state contact person (phone number)

_____

Emergency meeting place (address and phone)

_____

_____

_____

Emergency meeting place outside neighborhood (address)

_____

_____

_____

Evacuation Routes

_____

_____

_____

_____

_____

_____

35

Lost or Stranded Emergency Note

If you find this note please contact _____

_____

Date and Time departed: _____

_____

Direction of travel / destination: _____

_____

_____

_____

_____

Names and number of persons in group: _____

_____

_____

_____

_____

Special needs: _____

_____

_____

_____

Other: _____

_____

_____

_____

36

Emergency Utility Shutoff Locations

Gas: _____

_____

Electric: _____

_____

Water (Indoor Valve): _____

_____

Water (Outdoor Valve): _____

_____

Other: _____

_____

_____

_____

Notes

 Do not tear out this page.

(The other side is a form for your Utility locations)

Notes & Firestarter

(Tear out this page if you need paper to start a fire.)

I am so impressed with the cold steel knives,
I put them in my book for free.
Note: prices are only from their "Special Products" discount catalog
1-800-255-4716 for free catalog

Voyager Lockback Folding Knives

4" model #34MT
Tanto Point Aus-8 stainless
Lock withstands 60 pounds pressure
$48.99

Bush Ranger

model 37JS
Lifetime survival - camping knife
best bowie knife I know of
Very sharp, tough blade
will shave hair off your arm!
$85.99

Cold Steel Company
100 % satisfaction guaranteed
1-800-255-4716
www/coldsteel.com

Walkabout
Walking - Fighting Stick
model 91-WAC
made of tough hickory wood
$24.99

Boar Spear
model 95-BOA
Hunting or protection
very impressive
18 1/2 blade
$64.99
sheath $10.99

Cold Steel Company
100 % satisfaction guaranteed
1-800-255-4716
www/coldsteel.com

SCARS FIGHTING SYSTEM
Official System of the Navy Seals

DISCOUNT COUPON

Did you know that police estimate over 80% of the people in America will be assaulted in their lifetime?

What is the best fighting system in the world? Read this page then decide for yourself.

There is only **one** hand to hand combat system in the world that is the official course given to the Navy SEALS. That system is Jerry Peterson's SCARS course. Official government course # K-431-0096. You can call the United States Government Information Office and verify this for yourself. How good is the material in this course? No Navy SEAL has **ever** lost in combat when correctly using Jerry's system. It has also been proven to be very effective by regular citizens. This is not a course of fancy hard-to-learn martial arts moves. These are 100% proven, meat and potatoes fighting moves. These techniques are easy to learn. The videos explain everything. Many of these moves can be learned in one day! Now you can learn the same system as the fearless Navy SEALS in your home. You do not need to be strong, muscular, or athletic to fully master **any** of these techniques.

Please order now while this is on your mind.

I have been helping people get prepared long enough to know that if you put this off you will probably never order this course.

And this information could save your life someday. This is a **special discount price** available only through this coupon. The tapes come with a 90 day 100% **unconditional money back guarantee.**

ORDER FORM

Please send me:

( ) **Hostile Control Systems Package**
~~$99~~ Special Discount Price $80 plus $6 S&H (total $86 US$)

> **Tape I   Hostile Control Systems** excellent fighting techniques.
> **Tape II Hostile Control Systems** more advanced fighting techniques.
> **Companion Manual**, with 34 detailed instructions and human
>   target striking points.
> **Bonus: The Streetfighters Mindset** – Decision Points and Skill Tactics.
>   Might save your life!

( ) Enclosed is my check or money order (payable to **Direct Action Video**)

( ) Please charge my credit card: ( ) Visa ( ) Mastercard ( ) Discover ( ) AmEx

Credit Card No: _____

Expiration Date: _____

Name as it appears on card: _____

Name: (Print) _____

Address (Print) (no PO Box on int'l order) _____

_____

_____

City _____

Country _____

Telephone (____) _____
         (in case we have a question about your order)

This is a **mail in coupon only**. Not good with any other offers or discounts.
Cut out this coupon (**Xeroxes or copies of this coupon will NOT be accepted**).

Complete this page, cut out page, enclose payment and mail to:

Direct Action Video
5230 South 39th Street
Phoenix, AZ   85040

Notes

 Do not tear out this page.

(The other side is the Index)

INDEX

Supplies

It is very important that you buy your supplies now while you are thinking about them. If you don't buy them now (when they are on your mind), you may not get them at all. History has proven this to be true. Being prepared is the key to making it through disasters.

These companies sell disaster and emergency supplies. I know these to be good companies.

Major Surplus And Survival
(free catalog)
1-800-441-8855
435 West Alondra Blvd.
Garden, CA 90248

This company has fantastic bargains every month order thier free catalog right away.

Sportsman's Guide
(free catalog)
1-800-888-3006
Camping, Hunting, Fishing
closeout's, bargains
www.sportsmansguide.com